Lecture Notes in Applied Mechanics

Volume 5

Series Editor

Prof. Dr.-Ing. Friedrich Pfeiffer

Springer

Berlin
Heidelberg
New York
Barcelona
Hong Kong
London
Milan
Paris
Tokyo

Fourier BEM

Generalization of Boundary Element Methods by Fourier Transform

Fabian M. E. Duddeck

Springer

Dr.-Ing. Fabian M. E. Duddeck
Technische Universität München
Lehrstuhl für Baumechanik
Arcisstraße 21
80333 München
GERMANY

e-mail: fabian.duddeck@bv.tum.de

ISBN 3-540-43138-1 Springer Verlag Berlin Heidelberg New York

Library of Congress Cataloging-in-Publication Data

Duddeck, Fabian M.E., 1965-
Fourier BEM : generalization of boundary element methods by Fourier transform /
Fabian M.E. Duddeck.
p.cm. - (Lecture notes in applied mechanics ; v. 5)
Includes bibliographical references.
ISBN 3540431381 (hc : alk. paper)
1. Boundary element methods. 2. Fourier transformations. I. Title. II. Series.

TA 347.B69 D83 2002 2002075987
620'001'51535-dc21

Springer-Verlag Berlin Heidelberg New York
a member of BertelsmannSpringer Science+Business Media GmbH

http://www.springer.de

Printed in Germany

Cover design: design & production GmbH, Heidelberg
Typesetting: Digital data supplied by author

SPIN: 10864870 62/3020xv-5 4 3 2 1 0 - printed on acid-free paper

Fourier - BEM

Generalization of Boundary Element Methods by Fourier Transform

Dr.-Ing. Fabian M.E. Duddeck

March 10, 2002

Contents

Chapter 1

Introduction

Il y a plus de 50 ans que l'ingénieur Heaviside introduisit ses règles de calcul symbolique, dans un mémoire audacieux où des calculs mathématiques fort peu justifiés sont utilisés pour la solution de problèmes de physique. [...] Les ingénieurs utilisent [ce calcul symbolique] systématiquement, chacun avec sa conception personnelle, avec la conscience plus ou moins tranquille; c'est devenu une technique "qui n'est pas rigoureuse mais qui réussit bien".
L. Schwartz (1950), *Théorie des Distributions*, [Sch51][1]

1.1 Motivation

The research in recent decades has established the boundary element method (BEM) as a powerful tool in computational mechanics. Even dynamic and non-linear problems have been solved for fluid and solid mechanics. One of the remaining drawbacks is that the BEM as previously used is based on an explicit knowledge of fundamental solutions. In many engineering problems, e.g. anisotropic media, we do not know these fundamental solutions.

To overcome this drawback, an alternative BEM is presented here which is called here *Fourier-BEM* and is obtained by a spatial (and temporal)

[1]It is more than 50 years ago that the engineer Heaviside has introduced his symbolic calculus in an audacious report in which a mathematical calculus was used for the solution of physical problems which was not at all justified. [...] The engineers are using [this symbolic calculus] in a systematic way, everyone with his own personal conception, with a more or less tranquil conscience; it has become a technique "which is not rigorous but rather successful" [translated by F.D.].

Fourier transform of the traditional boundary integral equations (BIE). Only the Fourier transform of the fundamental solution is needed, which can be constructed, in contrast to the fundamental solution itself, *for all* linear and homogeneous differential operators by a simple matrix inversion of the transformed differential operator. Hence, the realm of applications of BEM can be extended remarkably.

1.2 Outline of Contents

In Chapter 2, the main aspects of the traditional approach are summarized. Attention here is focussed on direct Galerkin methods though we embed some aspects of other variants of BEM. The often rather mathematical idiom is "translated" into a representation understandable to engineers and visualized by the prototypical example of the mixed boundary value problem for the heat conduction (Poisson equation).

In Chapter 3, the traditional BEM is generalized by introducing a distributional notation developed here for the first time (distributions in the sense of Schwartz [Sch51]). Some special aspects are taken from the more recent theory of non-linear generalized functions introduced in the 80s by Colombeau [Col84, Col85, Col92]. The mathematical tools for the Fourier transformed BEM presented in the following part have been developed. Nevertheless, Chapter 3 may be skipped by readers not familiar with the theory of distributions although it contains the mathematically consistent and hence simpler description of jumps and singularities occurring in the context of the BEM.

In Chapter 4, the Fourier-BEM is presented. It introduces a new method for the assembly of stiffness matrices which works even if the fundamental solutions are not known. Hence, the approach presented here can be regarded as a generalization of the "classical" BEM. Many new fields of interest for an engineer are now accessible for analysis by a BEM-model.

Finally, several engineering examples are discussed to visualize the broad field of applications. Heat conduction, anisotropic elasticity, thin and thick plates, thermo- and poro-elasticity, static and dynamic elastic problems are investigated. In addition, first insights are given for applications to non-linear problems.

The integration of infinite and oscillating kernels is the main difficulty of the new approach. Hence, some ideas on optimized integration by wavelets are given in Chapter 11. This needs further investigation.

1.3 Biographical Context

Instead of giving an historical overview of BEM which can be found in the countless publications on BEM, e.g. the review papers [Ali97, Bes87, Bes97, BMP98], some biographical context is given here. It may serve to emphasize that BEM finds its place somewhere between applied mathematics and engineering mechanics. BEM is the product of a fruitful interaction between these two disciplines. Fourier-BEM generalizes the traditional symmetric Galerkin-BEM, hence we restrict the historical remarks on Fourier and Galerkin.

The life of Jean Baptiste Joseph Fourier:[2]

Joseph Fourier was born 21 March 1768 in Auxerre as the ninth of twelve children. His parents died early. In 1780 he started studying mathematics at the École Royale Militaire of Auxerre. In 1783 he received first prize for his work on Bossuet's Mécanique Générale. Fourier combined a religious vocation –

J.B.J. Fourier

B.G. Galerkin

in 1787 he entered the Benedictine abbey of St. Benoit-sur-Loire to train for the priesthood – with a mathematical orientation; he became a teacher at the École Royale Militaire of Auxerre in 1790. In 1793, he became involved in the revolutionary process and was imprisoned twice.

Later in 1794 Fourier was nominated to study at the École Normale in Paris, a teacher training school. Among his teachers were Lagrange, Laplace, and Monge. He was appointed to a position at the École Centrale des Travaux Publiques, which was under the direction of Lazare, Carnot and Monge, and

[2] Paraphrase of an electronic article written by J.J. O'Connor and E.F. Robertson; http://www-history.mcs.st-andrews.ac.uk/history/Mathematicians/Fourier.html.

soon to be renamed École Polytechnique. In 1797 he succeeded Lagrange in being appointed to the chair of analysis and mechanics.

In 1798 Fourier joined Napoleon's army as scientific adviser in its invasion of Egypt. He helped to establish educational facilities in Egypt, carried out archaeological explorations, and became secretary of the newly founded Cairo Institute. Returning to France in 1801, Fourier resumed his post as Professor of Analysis at the École Polytechnique.

On Napoleons initiative, Fourier was nominated as Prefect in Grenoble. In this position he was responsible for the drainage of the swamps of Bourgoin, and for the construction of a new highway from Grenoble to Turin.

During his time in Grenoble Fourier worked on the mathematical theory of heat. He submitted *On the Propagation of Heat in Solid Bodies* to the Paris Institute on 21 December 1807 where it was criticized by the scientific committee (Lagrange, Laplace, Monge and Lacroix). The first objection, made by Lagrange and Laplace in 1808, was to Fourier's expansions of functions as trigonometrical series, what we now call Fourier series. Further clarification by Fourier still failed to convince them. The second objection was made by Biot against Fourier's derivation of the equations of transfer of heat. Fourier had not made reference to Biot's 1804 paper on this topic although Biot's paper is certainly incorrect. Laplace, and later Poisson, had similar objections.

Fourier was elected to the Académie des Sciences in 1817. In 1822 Delambre died, who was secretary to the mathematical section of the Académie des Sciences, and Fourier together with Biot and Arago applied for the post. After Arago withdrew, the election gave Fourier an easy win. Shortly after Fourier became secretary, the Academy published his prize winning essay *Théorie analytique de la chaleur* in 1822.

During Fourier's last eight years in Paris he resumed his mathematical researches and published a number of papers, some in pure mathematics others on applied mathematical topics. However, his scientific life was not without problems, since his theory of heat still provoked controversy. Biot claimed priority over Fourier, a claim which Fourier had little difficulty refuting. Poisson, however, attacked both Fourier's mathematical techniques and claimed to have an alternative theory. Fourier wrote *Historical Précis* as a reply to these claims but, although the work was shown to various mathematicians, it was never published. Fourier died 16 May 1830 in Paris.

The life of Boris Grigorievich Galerkin:[3]
Galerkin was born 4 March 1871 in Polotsk, Belarus. From 1893 to 1899, he studied engineering and mathematics at the Petersburg Technological Institute (PTI). He started teaching in 1909 while working as an engineer, first at the Kharkov Locomotive Plant and later as an engineering manager at the Northern Mechanical and Boiler Plant. His first publications range form practical engineering aspects for bridges and frames to theoretical mathematical areas, namely the method of approximate integration of differential equations known today as the *Galerkin method.*

In 1920 Galerkin was promoted to Head of Structural Mechanics at the PTI. By this time he also held two chairs, one in elasticity at the Leningrad Institute of Communications Engineers and one in structural mechanics at Leningrad University. In 1921 the St Petersburg Mathematical Society was reopened (it had closed in 1917 due to the Russian Revolution) as the Petrograd Physical and Mathematical Society. Galerkin played a major role in the Society along with Steklov, Sergi Bernstein, Friedmann and others.

Galerkin was a consultant in the planning and building of many of the Soviet Union's largest hydrostations. In 1929, in connection with the building of the Dnjepr dam and hydroelectric station, Galerkin investigated stresses in dams and breast walls with trapezoidal profile.

Other works for which Galerkin is famous are his studies of thin elastic plates. His major monograph on this topic *Thin Elastic Plates* was published in 1937. From 1940 until his death 1945, Galerkin was head of the Institute of Mechanics of the Soviet Academy of Sciences.

[3]Paraphrase of an electronic article written by J.J. O'Connor and E.F. Robertson; http://www-history.mcs.st-andrews.ac.uk/history/Mathematicians/Galerkin.html.

Chapter 2

Traditional BEM

Enrico Betti

2.1 Introduction

BEM makes possible solutions to differential problems in complex geometries. Values, e.g. the displacements and volume forces, in the interior of an n-dimensional domain $\Omega \subset \mathbb{R}^n, n = 2, 3$ (Fig. 2.1) are related to exterior quantities like the displacements and the tractions at the boundary $\partial\Omega$.

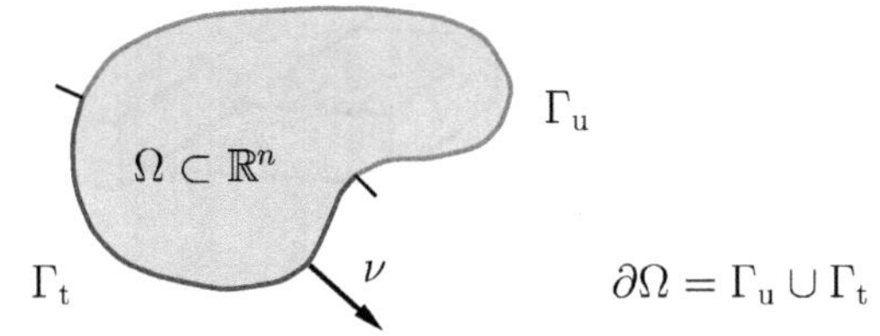

Figure 2.1: *Domain Ω with Dirichlet and Neumann boundary.*

The known and unknown boundary values are approximated by polynomial trial functions, whereas the interior remains undiscretized. This results, in contrast to FEM, in a relatively low number of unknowns. An n-dimensional problem is reduced to one of the order $n-1$. Then, an algebraic system of equations is obtained by a procedure of weighted residuals. However, the underlying matrices are in general asymmetric and totally populated, as a result of the interrelation of all boundary values.

2.2 Boundary Integral Equations (BIE)

2.2.1 Reciprocity relation

Geometrical restrictions

The BEM can be applied to problems in arbitrary geometries. The only condition needed is that the boundary should be *Lipschitzian.* The Lipschitz-condition is required for the Gauß theorem which is the basic theorem of all BEMs. A unit outer normal vector ν should be defined everywhere on the boundary $\partial\Omega$ with the exception of single points. Corners and edges are allowed.

Most of the cases which occur in engineering mechanics are problems posed in such Lipschitz-domains, although there are some important cases, where

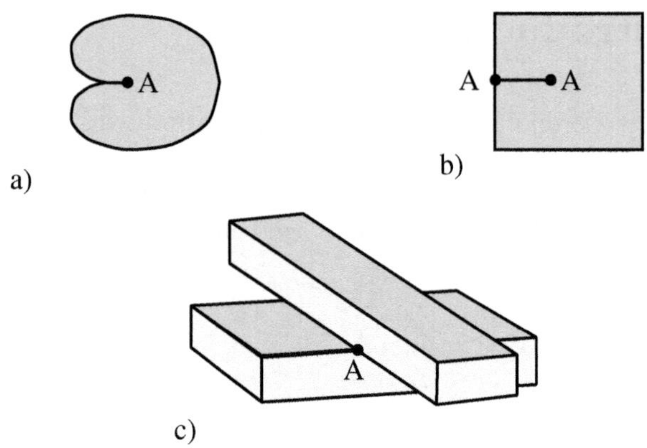

Figure 2.2: *Non-Lipschitz domains [McL00].*

the Lipschitz condition is not fulfilled (Fig. 2.2), cf. McLean [McL00], pp. 89. In part a) of Fig. 2.2, the cusp at point A is too sharp, i.e. the boundaries approach one another faster than any polynomial. The crack with zero crack width in Fig. 2.2b) is not Lipschitzian because the domain lies on both sides of the boundary. And in Fig. 2.2c) we cannot represent the boundary in the vicinity of the point A as a graph of a function. In all these cases, we may divide the domain into substructures. Especially for cracks with zero crackwidth one may find several modifications of the BEM in the literature to overcome this restriction, cf. the review of Aliabadi [Ali97].

Heat conduction as a paradigmatic example

The basic principles of traditional BEM are presented for the paradigmatic example of the n-dimensional stationary heat conduction which is described by the *Poisson equation*

$$\begin{aligned} \Delta u(x) &= -f(x), && x \in \Omega \subset \mathbb{R}^n,\ \Delta = \sum_{k=1}^{n} \partial^2/\partial x_k^2; \qquad (2.1)\\ u(x) &= u_\Gamma(x), && x \in \Gamma_\mathrm{u} \subset \partial\Omega;\\ t(x) &= t_\Gamma(x), && x \in \Gamma_\mathrm{t} \subset \partial\Omega. \end{aligned}$$

Δ is the Laplace operator, u is the unknown quantity (temperature) and f denotes known volume sources in Ω. The flux at the boundary is

$$t = \mathcal{A}_\mathrm{t} u = -\partial_\nu u = -\nu \cdot \nabla u. \qquad (2.2)$$

∇, ν are the gradient and the outer unit normal; $\mathcal{A}_\mathrm{t} = -\nu \cdot \nabla$ is the boundary operator. The partial derivatives $\partial/\partial x_k$ will be abbreviated by ∂_k. x denotes the n-dimensional vector and $\mathrm{d}x$ is the short form for $\mathrm{d}x_1\,\mathrm{d}x_2$ in the two-dimensional case and $\mathrm{d}x_1\,\mathrm{d}x_2\,\mathrm{d}x_3$ in three dimensions. To obtain a well posed problem, half of the boundary data, either u on Γ_u or t on Γ_t, should be defined by boundary conditions, i.e. $\Gamma_\mathrm{u} \cup \Gamma_\mathrm{t} = \partial\Omega$.

The weak form of the differential equation (reciprocity)

Most of the numerical methods are based on a *weak form* equivalent to the strong form (2.1). They are obtained by the multiplication by a second function and by the integration over the domain (weighting). In the engineering literature this is known as the principle of virtual work. The shifting of the differential operator to the weighting function by partial integration leads to additional boundary terms. Thus, the basic weak form for BEM, the *reciprocity relation* is obtained

$$\int_\Omega u\,\Delta v\,\mathrm{d}\Omega \;=\; \int_\Omega \Delta u\,v\,\mathrm{d}\Omega - \int_\Gamma \left(\frac{\partial u}{\partial \nu} v - u\frac{\partial v}{\partial \nu}\right)\mathrm{d}\Gamma. \qquad (2.3)$$

It is also known as *Betti's theorem* [1] and is equivalent to *Green's formula*[2]. If $v = u_{II}, u = u_I$ are regarded as two different states (real and virtual) and if we introduce the notations $\Delta u_I = -f_I; \Delta u_{II} = -f_{II}; t_I = -\partial_\nu u_I; t_{II} = -\partial_\nu u_{II}$, then the reciprocal character of equation (2.3) is obvious

$$\int_\Omega u_I\,f_{II}\,\mathrm{d}\Omega \;=\; \int_\Omega f_I\,u_{II}\,\mathrm{d}\Omega - \int_\Gamma (t_I\,u_{II} - u_I\,t_{II})\,\mathrm{d}\Gamma.$$

[1] E. Betti *Il Nuovo Ciemento* **7–8** (1872).

[2] G. Green: *An Essay on the Application of Mathematical Analysis to the Theories of Electricity and Magnetism.* Nottingham (1828).

2.2.2 Inversion of the differential operator

The fundamental solution as inverse of the differential operator

From (2.3), the basic integral equations for the BEM are constructed by an inversion of the differential operator Δ. This can be achieved by the con-

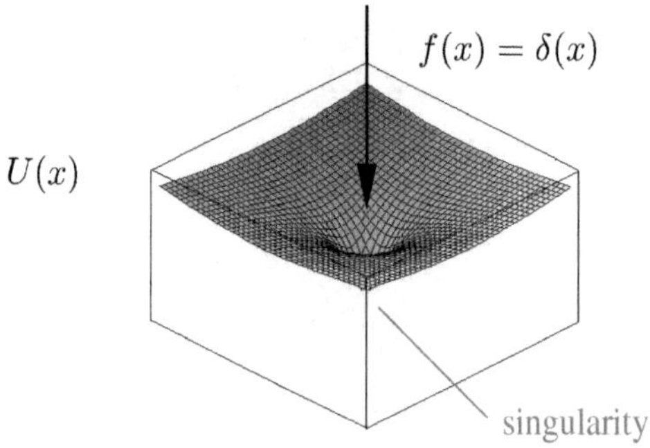

Figure 2.3: *The fundamental solution U as response to a single source δ.*

volution with the *fundamental solution* $U(x)$, i.e. we have to replace v by $U(x-y)$, where $U(x-y)$ is the response of the infinite medium to a single source $f = \delta$ at $x = y$ (Fig. 2.3)

$$\Delta U(x-y) = -\delta(x-y), \qquad x, y \in \mathbb{R}^n. \tag{2.4}$$

For the Laplacian Δ, we obtain for two- or three-dimensional problems [DL00]

$$U(x) = -\frac{1}{2\pi}\ln|x| - \frac{C_0}{2\pi}, \qquad x \in \mathbb{R}^2; \tag{2.5}$$

$$U(x) = -\frac{1}{4\pi}\frac{1}{|x|}, \qquad x \in \mathbb{R}^3. \tag{2.6}$$

The constant C_0 in (2.5) is arbitrary. It represents rigid modes which are homogeneous solutions of Δ. The fundamental solutions inherit their singular character from the Dirac-distribution. Unfortunately, these analytic formulas for the fundamental solutions can only be found for simple differential operators, cf. [GM92, Kyt96, Kög00]. Nevertheless, as long as the coefficients of the differential operator are constant, the existence of a fundamental solution can always be assured.

Somigliana identity

The insertion of the fundamental solution $U(x-y)$ into (2.3) leads to

$$\begin{aligned}\int_\Omega u(y)\,\Delta U(x-y)\,\mathrm{d}\Omega_y &= \int_\Omega \Delta u(y)\,U(x-y)\,\mathrm{d}\Omega_y \\ &+ \int_{\Gamma_y} [u(y)\,T(x-y) - t(y)\,U(x-y)]\,\mathrm{d}\Gamma_y,\end{aligned} \tag{2.7}$$

with $T = \mathcal{A}_\mathrm{t}U$. On the left-hand side, we substitute

$$\Delta U(x-y) = -\delta(x-y), \qquad x, y \in \mathbb{R}^n. \tag{2.8}$$

The Dirac-distribution δ has in Ω the convolution property

$$\int_\Omega u(y)\delta(x-y)\,\mathrm{d}\Omega_y = u(x), \qquad x \in \Omega. \tag{2.9}$$

The relation (2.9) is not valid for boundary points, i.e. $x \notin \partial\Omega$. Together with (2.1) we obtain from (2.7) to (2.9) *Somigliana's identity*[3] for interior points $x \in \Omega$:

$$\begin{aligned}u(x) &= \int_\Omega f(y)\,U(x-y)\,\mathrm{d}\Omega_y \\ &+ \int_{\Gamma_y} [t(y)\,U(x-y) - u(y)\,T(x-y)]\,\mathrm{d}\Gamma_y.\end{aligned} \tag{2.10}$$

2.2.3 Galerkin and collocation BIE

Boundary integral equations (BIE)

The transition to boundary integral equations (BIE) demands a limit process $x \to \partial\Omega$, cf. for example [Bon99]. Hereby the left-hand side of (2.10) is modified by a factor $\kappa(x) \in [0,1]$ due to the special boundary relation

$$\int_\Omega u(y)\,\delta(x-y)\,\mathrm{d}\Omega_y = \kappa(x)u(x), \qquad x, y \in \partial\Omega. \tag{2.11}$$

We obtain the *boundary integral equation*

$$\begin{aligned}\kappa(x)u(x) &= \int_\Omega f(y)\,U(x-y)\,\mathrm{d}\Omega_y \\ &+ \int_{\Gamma_y} [t(y)\,U(x-y) - u(y)\,T(x-y)]\,\mathrm{d}\Gamma_y.\end{aligned} \tag{2.12}$$

For smooth boundaries, $\kappa(x)$ is equal to 1/2. Equation (2.11) is justified by a rigorous distributional argumentation in Chapter 3.2.

[3] C. Somigliana: Sopra l'equilibrio di un corpo elastico isotrope. *Il Nuovo Ciemento* **17–19** (1886).

Approximation of the boundary quantities

The known and unknown boundary quantities u, t are approximated by a sum of piecewise polynomial *trial functions* $\phi_{\mathrm{u}}^i, \phi_{\mathrm{t}}^i$ with the coefficients $\mathrm{u}^i, \mathrm{t}^i$ (Fig. 2.4)

$$u(x) \approx \sum_i^{N_{\mathrm{u}}} \mathrm{u}^i \phi_{\mathrm{u}}^i(x), \qquad t(x) \approx \sum_i^{N_{\mathrm{t}}} \mathrm{t}^i \phi_{\mathrm{t}}^i(x). \tag{2.13}$$

For convergence reasons, the trial functions for the temperature u should be at least linear whereas for the flux t, it is sufficient to take constant trial functions, cf. for example [Bon95a][4].

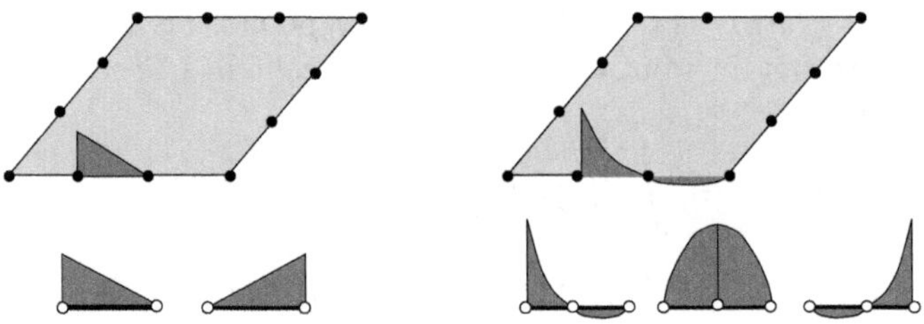

Figure 2.4: *The linear and quadratic trial functions* $\phi_{\mathrm{u,t}}^i$.

Galerkin boundary integral equations

An additional weighting with one of the trial functions for the flux (2.13), e.g. [Hol92, BMP98], leads to

$$\begin{aligned} \int_{\Gamma_x} \phi_{\mathrm{t}}^j(x) \kappa(x) u(x) \, \mathrm{d}\Gamma_x = & \int_{\Gamma_x} \phi_{\mathrm{t}}^j(x) \int_{\Omega} f(y) \, U(x-y) \, \mathrm{d}\Omega_y \, \mathrm{d}\Gamma_x \\ & + \int_{\Gamma_x} \phi_{\mathrm{t}}^j(x) \int_{\Gamma_y} [t(y) \, U(x-y) - u(y) \, T(x-y)] \, \mathrm{d}\Gamma_y \, \mathrm{d}\Gamma_x. \end{aligned} \tag{2.14}$$

Here the trial functions of the flux are chosen because t is the dual variable to u (dual in the sense that the two variables are the two factors in the scalar product of work) and the outer integrals have to be work terms. Introducing

[4]The Galerkin approach weakens these continuity requirements; only $C^{1,\alpha}$ is necessary for the hypersingular integrals [KRR92, MR96].

(2.13) in (2.14) the discretized *Galerkin BIE* is obtained :

$$\int_{\Gamma_x} \phi_{\mathrm{t}}^j(x)\kappa(x)u(x)\,\mathrm{d}\Gamma_x = \int_{\Gamma_x} \phi_{\mathrm{t}}^j(x) \int_{\Omega} f(y)\,U(x-y)\,\mathrm{d}\Omega_y\,\mathrm{d}\Gamma_x \tag{2.15}$$
$$+\sum_i^{N_{\mathrm{t}}} \mathrm{t}^i \int_{\Gamma_x} \phi_{\mathrm{t}}^j(x) \int_{\Gamma_y} \phi_{\mathrm{t}}^i(y)\,U(x-y)\,\mathrm{d}\Gamma_y\,\mathrm{d}\Gamma_x$$
$$-\sum_i^{N_{\mathrm{u}}} \mathrm{u}^i \int_{\Gamma_x} \phi_{\mathrm{t}}^j(x) \int_{\Gamma_y} \phi_{\mathrm{u}}^i(y)\,T(x-y)\,\mathrm{d}\Gamma_y\,\mathrm{d}\Gamma_x.$$

Collocation boundary integral equations
To avoid double integration in (2.15), Dirac-distributions $\phi_{\mathrm{t}}^j(x) = \delta(x-\xi^j)$ can be chosen as test functions with $\xi^j \in \partial\Omega$. Due to the sifting property $\int_{\mathbb{R}^n} u(x)\delta(x-\xi^j)\,\mathrm{d}x = u(\xi^j)$ this leads to the discretized *collocation BIE*

$$\kappa(\xi^j)u(\xi^j) = \int_{\Omega} f(y)\,U(\xi^j-y)\,\mathrm{d}\Omega_y \tag{2.16}$$
$$+ \sum_i^{N_{\mathrm{t}}} \mathrm{t}^i \int_{\Gamma_y} \phi_{\mathrm{t}}^i(y)\,U(\xi^j-y)\,\mathrm{d}\Gamma_y - \sum_i^{N_{\mathrm{u}}} \mathrm{u}^i \int_{\Gamma_y} \phi_{\mathrm{u}}^i(y)\,T(\xi^j-y)\,\mathrm{d}\Gamma_y.$$

In contrast to the Galerkin approach, the collocation method leads to non-symmetric matrices, i.e., the system looses its self-adjoint property, cf. [AS97]. It is not always clear where to put the collocation points and the theoretical reasoning, e.g. convergence proofs [MP96], is much more difficult than for the Galerkin method. The coupling with the FEM is not as straightforward as for the Galerkin-BEM.

2.3 Complete System of BIE

2.3.1 Differentiation of BIE

Somigliana identity for the flux
For the symmetric Galerkin method, an additional BIE is needed for the flux which can be obtained by applying the adjoint boundary operator $\mathcal{A}_{\mathrm{t}}^j = -\partial_\nu$ on (2.10), i.e. we have to differentiate with respect to the normal of the

j-th test function

$$\mathcal{A}_{\mathrm{t}}^{j} u(x) = \mathcal{A}_{\mathrm{t}}^{j} \int_{\Omega} f(y)\, U(x-y)\, \mathrm{d}\Omega_y \tag{2.17}$$
$$+\mathcal{A}_{\mathrm{t}}^{j} \sum_{i}^{N_{\mathrm{t}}} \mathrm{t}^i \int_{\Gamma_y} \phi_{\mathrm{t}}^i(y)\, U(x-y)\, \mathrm{d}\Gamma_y - \mathcal{A}_{\mathrm{t}}^{j} \sum_{i}^{N_{\mathrm{u}}} \mathrm{u}^i \int_{\Gamma_y} \phi_{\mathrm{u}}^i(y)\, \mathcal{A}_{\mathrm{t}}^{i} U(x-y)\, \mathrm{d}\Gamma_y,$$

where T was replaced by $\mathcal{A}_{\mathrm{t}}^{i} U$. The differentiation and integration can be interchanged because of the following equality which is valid for all convolutional products even when the factors have severe singularities, cf. [Dud99a][5],

$$\mathcal{A}_{\mathrm{t}}^{j}(u * v) \;=\; (\mathcal{A}_{\mathrm{t}}^{j} u) * v = u * (\mathcal{A}_{\mathrm{t}}^{j} v). \tag{2.18}$$

Thus, we obtain *Somigliana's identity for the flux* for interior points $x \in \Omega$

$$t(x) = \int_{\Omega} f(y)\, \mathcal{A}_{\mathrm{t}}^{j}\, U(x-y)\, \mathrm{d}\Omega_y \tag{2.19}$$
$$+\sum_{i}^{N_{\mathrm{t}}} \mathrm{t}^i \int_{\Gamma_y} \phi_{\mathrm{t}}^i(y)\, \mathcal{A}_{\mathrm{t}}^{j} U(x-y)\, \mathrm{d}\Gamma_y - \sum_{i}^{N_{\mathrm{u}}} \mathrm{u}^i \int_{\Gamma_y} \phi_{\mathrm{u}}^i(y)\, \mathcal{A}_{\mathrm{t}}^{j} \mathcal{A}_{\mathrm{t}}^{i} U(x-y)\, \mathrm{d}\Gamma_y.$$

Galerkin BIE for flux

To get the BIE for boundary points $x \in \partial\Omega$, u has to be replaced by κu on the right-hand side of (2.17). The differentiation of the product is then

$$\mathcal{A}_{\mathrm{t}}^{j}\{\kappa(x) u(x)\} \;=\; \kappa(x) \mathcal{A}_{\mathrm{t}}^{j} u(x) + u(x) \mathcal{A}_{\mathrm{t}}^{j} \kappa(x). \tag{2.20}$$

The value $\kappa(x)$ gives the "percentage" of the Dirac-distribution $\delta(x)$ relevant for the domain. When weighted with a test function ϕ_{u}^{j} the second term becomes infinite (the non-integrable singularity cancels with that originating from the hyper singular part on the right-hand side as shown later). The *Galerkin BIE for flux* is formally :

$$\int_{\Gamma_x} \phi_{\mathrm{u}}^{j}(x) \mathcal{A}_{\mathrm{t}}^{j}\{\kappa(x) u(x)\}\, \mathrm{d}\Gamma_x = \int_{\Gamma_x} \phi_{\mathrm{u}}^{j}(x) \int_{\Omega} f(y)\, \mathcal{A}_{\mathrm{t}}^{j} U(x-y)\, \mathrm{d}\Omega_y\, \mathrm{d}\Gamma_x \tag{2.21}$$
$$+\int_{\Gamma_x} \phi_{\mathrm{u}}^{j}(x) \sum_{i}^{N_{\mathrm{t}}} \mathrm{t}^i \int_{\Gamma_y} \phi_{\mathrm{t}}^i(y) \mathcal{A}_{\mathrm{t}}^{j} U(x-y)\, \mathrm{d}\Gamma_y\, \mathrm{d}\Gamma_x$$
$$-\int_{\Gamma_x} \phi_{\mathrm{u}}^{j}(x) \sum_{i}^{N_{\mathrm{u}}} \mathrm{u}^i \int_{\Gamma_y} \phi_{\mathrm{u}}^i(y) \mathcal{A}_{\mathrm{t}}^{j} \mathcal{A}_{\mathrm{t}}^{i} U(x-y)\, \mathrm{d}\Gamma_y\, \mathrm{d}\Gamma_x.$$

The flux test functions ϕ_{t}^{j} were replaced by those of the temperature ϕ_{u}^{j} to preserve duality. Otherwise the matrices will be asymmetric.

[5] The only condition is : the support of one of the two convolutional factors has to be compact, [Hör90], p. 103. All trial functions have compact support.

Differentiation of the fundamental solution
In the BIEs, the following differentiations of U are needed :

$$\partial_k U = -\frac{1}{2\pi}\frac{x_k}{|x|^2}, \qquad x \in \mathbb{R}^2; \tag{2.22}$$

$$\partial_k U = -\frac{1}{4\pi}\frac{x_k}{|x|^3}, \qquad x \in \mathbb{R}^3. \tag{2.23}$$

Similarly for the second derivatives in the $\mathbb{R}^2$:

$$[\partial_{kl} U] = \frac{1}{2\pi|x|^4}\begin{bmatrix} x_1^2 - x_2^2 & 2x_1x_2 \\ 2x_1x_2 & x_2^2 - x_1^2 \end{bmatrix} \tag{2.24}$$

and in the $\mathbb{R}^3$:

$$[\partial_{kl} U] = \frac{-1}{4\pi|x|^5}\begin{bmatrix} 2x_1^2 - x_2^2 - x_3^2 & 3x_1x_2 & 3x_1x_3 \\ 3x_2x_1 & 2x_2^2 - x_3^2 - x_1^2 & 3x_2x_3 \\ 3x_3x_1 & 3x_3x_2 & 2x_3^2 - x_1^2 - x_2^2 \end{bmatrix} \tag{2.25}$$

2.3.2 Symmetric system of BIEs

The Galerkin BIEs (2.15, 2.21) lead to the *algebraic system of BIEs*

$$\text{from (2.15)}: \quad \sum_i K_{\mathrm{u}}^{ji}\mathrm{u}^i = F_{\mathrm{u}}^j + \sum_i H_{\mathrm{u}}^{ji}\mathrm{t}^i - \sum_i G_{\mathrm{u}}^{ji}\mathrm{u}^i; \tag{2.26}$$

$$\text{from (2.21)}: \quad \sum_i K_{\mathrm{t}}^{ji}\mathrm{t}^i = F_{\mathrm{t}}^j + \sum_i H_{\mathrm{t}}^{ji}\mathrm{t}^i - \sum_i G_{\mathrm{t}}^{ji}\mathrm{u}^i,$$

where known and unknown boundary values are still not distinguished. The vector and matrices are defined as follows :

$$F_{\mathrm{u}}^j := \int_{\Gamma_x} \phi_{\mathrm{t}}^j(x) \int_\Omega f(y)\, U(x-y)\, \mathrm{d}\Omega_y\, \mathrm{d}\Gamma_x; \tag{2.27}$$

$$H_{\mathrm{u}}^{ji} := \int_{\Gamma_x} \phi_{\mathrm{t}}^j(x) \int_{\Gamma_y} \phi_{\mathrm{t}}^i(y)\, U(x-y)\, \mathrm{d}\Gamma_y\, \mathrm{d}\Gamma_x;$$

$$G_{\mathrm{u}}^{ji} := \int_{\Gamma_x} \phi_{\mathrm{t}}^j(x) \int_{\Gamma_y} \phi_{\mathrm{u}}^i(y)\, \mathcal{A}_{\mathrm{t}}^i U(x-y)\, \mathrm{d}\Gamma_y\, \mathrm{d}\Gamma_x;$$

$$K_{\mathrm{u}}^{ji} := \int_{\Gamma_x} \phi_{\mathrm{t}}^j(x)\, \kappa(x) \phi_{\mathrm{u}}^i(x)\, \mathrm{d}\Gamma_x,$$

and

$$
\begin{aligned}
F_{\mathrm{t}}^{j} &:= \int_{\Gamma_x} \phi_{\mathrm{u}}^{j}(x) \int_{\Omega} f(y)\, \mathcal{A}_{\mathrm{t}}^{j} U(x-y)\, \mathrm{d}\Omega_y\, \mathrm{d}\Gamma_x; \\
H_{\mathrm{t}}^{ji} &:= \int_{\Gamma_x} \phi_{\mathrm{u}}^{j}(x) \int_{\Gamma_y} \phi_{\mathrm{t}}^{i}(y)\, \mathcal{A}_{\mathrm{t}}^{j} U(x-y)\, \mathrm{d}\Gamma_y\, \mathrm{d}\Gamma_x; \\
G_{\mathrm{t}}^{ji} &:= \int_{\Gamma_x} \phi_{\mathrm{u}}^{j}(x) \int_{\Gamma_y} \phi_{\mathrm{u}}^{i}(y)\, \mathcal{A}_{\mathrm{t}}^{j} \mathcal{A}_{\mathrm{t}}^{i} U(x-y)\, \mathrm{d}\Gamma_y\, \mathrm{d}\Gamma_x; \\
K_{\mathrm{t}}^{ji} &:= \int_{\Gamma_x} \phi_{\mathrm{u}}^{j}(x)\, (\kappa \phi_{\mathrm{t}}^{i} + \phi_{\mathrm{u}}^{i} \mathcal{A}_{\mathrm{t}}^{j} \kappa)\, \mathrm{d}\Gamma_x.
\end{aligned}
\tag{2.28}
$$

By reordering, where known boundary values are concatenated with the force vector, the final matrix system comparable to that of the FEM is obtained

$$
\sum_i A^{ji} X^i = Y^j \tag{2.29}
$$

for the evaluation of the unknown boundary quantities X. The system matrix A is fully populated and symmetric.

Chapter 3

Distributional BEM

Oliver Heaviside

3.1 Distributional Context

To obtain the Fourier transform of the BIE derived above, all quantities have to be extended from Ω to $\mathbb{R}^n$ (the Fourier transformation is defined on $\mathbb{R}^n$ and not on Ω). Formally, this can be done by defining a cutoff distribution χ which is simply one in the interior of Ω and zero outside. Then all quantities are multiplied by χ and finally transformed into Fourier space (windowed Fourier transform). Mathematically this extension and transformation is justified only in the frame of the theory of distributions.

This theory is fundamental for integral equations. It simplifies the derivation of the reciprocity relations, the correct treatment of discontinuities and singularities, and the evaluation of the free terms. The Fourier transform without its distributional extension is not complete enough to treat the simplest problems. Divergent integrals are declared.

The main advantage of the theory of distributions is that it reestablishes differentiation as a simple and consistent procedure, all quantities are differentiable even if they exhibit severe singularities or jumps, cf. Dieudonné [Die78].

Historically the distribution theory was first initiated heuristically by the engineer Heaviside[1]. Later in the forties, Sobolev[2] introduced the first mathematically rigorous concept which was finally established by Schwartz in 1950

[1] O. Heaviside: On operators in physical mathematics. *Proc. Roy. Soc. A* **52**, 504–529 (1893).

[2] S.L. Sobolev, Sur un théorème d'analyse fonctionnelle (Russian; French summary). *Mat. Sb.* **46** (4), 471–497 (1938), and *Amer. Math. Soc. Transl.* **34**(2), 39–68 (1963).

[Sch51]. Further developments can be found in the work of Gel'fand & Shilov [GS64]. Here, reference is given mostly to the standard book on linear differential operators by Hörmander [Hör90].

S.L. Sobolev (1908 – 89) L. Schwartz (∗1915) I.M. Gel'fand (∗1913)

Figure 3.1: *Founders of the theory of distributions*

Some additional ideas are taken from the more recent literature on non-linear generalized functions initiated by Colombeau in his books [Col84, Col85, Col92]. Recent results can be found in [Bia90, Obe92, OR94, Boi96]. Even the product of two Dirac-distributions $\delta(x)$ is now defined.

The reader not acquainted with distribution theory may switch directly to the next chapter keeping in mind that all quantities are extended from Ω to $\mathbb{R}^n$ and that Fourier transformation can now be applied to the BIE established before.

3.2 Distributional Description of the Domain

3.2.1 The cutoff - distribution

The cutoff distribution χ as a description of the domain Ω

The unknown quantity u extends only over Ω and not over $\mathbb{R}^n$. It may jump across the boundary $\partial\Omega$. In the theory of distributions, this is described by a multiplication of $u \in C^\infty(\mathbb{R}^n)$ with the *cutoff distribution* χ (characteristic distribution) of the domain Ω

$$u(x) \to \chi(x)u(x), \qquad \chi(x) := \begin{cases} 1 & \dots \quad x \in \Omega \\ \kappa(x) & \dots \quad x \in \partial\Omega \\ 0 & \dots \quad x \notin \bar{\Omega} = \Omega \cup \partial\Omega. \end{cases} \tag{3.1}$$

χ equals one in the interior and vanishes outside the domain Ω and can be expressed for smooth boundaries by a generalized multi-dimensional Heaviside-distribution, cf. [Con74, GS64, Hör90],

$$\chi(x) = \mathrm{H}(\psi(x)), \qquad x \in \mathbb{R}^n. \tag{3.2}$$

ψ describes as a hypersurface the boundary $\partial\Omega$. Domains with non-smooth boundaries are expressed by products and sums of these Heaviside-distributions [Dud97b].

Example 3.2.1 cutoff distribution for the semicircle
The cutoff distribution of a semicircle in $\mathbb{R}^2$ of unit radius is (Fig. 3.2)

$$\chi(x) \quad = \quad \mathrm{H}(x_1)\mathrm{H}(1 - x_1^2 - x_2^2). \tag{3.3}$$

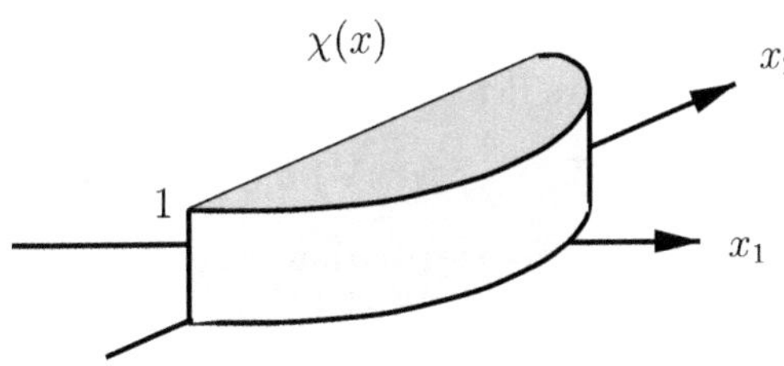

Figure 3.2: *The cutoff distribution χ of the semicircle*

The edge and corner term κ of the cutoff distribution
The value $\kappa(x)$ of $\chi(x), x \in \partial\Omega$ directly at the boundary $\partial\Omega$, i.e. at edges and corners, can be determined by a convolution with the Dirac-distribution

$$\kappa(x) = \chi(x)|_{x\in\partial\Omega} = \int_{\mathbb{R}^n} \chi(y)\,\delta(x-y)\,\mathrm{d}y, \qquad y \in \mathbb{R}^n, x \in \partial\Omega. \tag{3.4}$$

This convolution is well defined but often difficult to compute. Hence, we refer to the non-linear theory of distributions, cf. [Col84, Col85, Col92, Bia90, Obe92, OR94, Boi96], where the Dirac-distribution is often replaced by a Dirac-convergent sequence. We define for example:

$$\delta(x) \quad = \quad \lim_{k\to\infty} \phi_k(x) = \lim_{k\to\infty} \frac{k^n}{\sqrt{\pi^n}}\,\mathrm{e}^{-k^2|x|^2}, \qquad x \in \mathbb{R}^n. \tag{3.5}$$

The convolution (3.4) is now evaluated by

$$\kappa(x) = \int_{\mathbb{R}^n} \chi(y)\, \delta(x-y)\, \mathrm{d}y = \lim_{k\to\infty} \int_{\mathbb{R}^n} \chi(y)\phi_k(x-y)\, \mathrm{d}y. \tag{3.6}$$

For curved boundaries, we approximate χ locally by a tangential cutoff distribution.

3.2.2 The normal vector

In the distributional context, the normal vector ν can be generally defined by the gradient of the hypersurface ψ. We have for smooth boundaries

$$\nabla\chi = \nabla \mathrm{H}(\psi) = \delta(\psi)\nabla\psi = -\delta(\psi)\,\nu. \tag{3.7}$$

The Heaviside-distribution is dilational invariant

$$\mathrm{H}(\psi) = \mathrm{H}(a\psi), \qquad a = a(x) > 0, \tag{3.8}$$

which is also valid for the gradient

$$\nabla \mathrm{H}(\psi) = \delta(\psi)\nabla\psi = \delta(a\psi)\nabla\{|a|\psi\} = \nabla \mathrm{H}(a\psi). \tag{3.9}$$

although we have for the Dirac-distribution, cf. [GS64], p. 213,

$$\delta(\psi) = |a|\delta(a\psi); \tag{3.10}$$

$$\partial^k\delta(\psi) = |a|^{k+1}\partial^k\delta(a\psi). \tag{3.11}$$

Hence, in the distributional representation there is no need for normalization. Examples are given in the Appendix C.1.

3.3 Distributional BIE

3.3.1 Distributional reciprocity

Distributional differentiation of the product χu

The generalized differentiation of the product χu leads to terms in the interior of Ω and to additional boundary terms[3]. In our example of the Laplacian, we obtain

$$\Delta\{\chi u\} = \chi\Delta u + 2\nabla\chi\cdot\nabla u + u\Delta\chi. \tag{3.12}$$

[3]Leibniz' formula for the differentiation of products remains valid for distributions, Hörmander [Hör90], equation (3.1.4).

An additional weighting with v leads to

$$\int_{\mathbb{R}^n} v\Delta\{\chi u\}\,\mathrm{d}y = \int_{\mathbb{R}^n} v\chi\Delta u\,\mathrm{d}y + 2\int_{\mathbb{R}^n} v\nabla\chi\cdot\nabla u\,\mathrm{d}y + \int_{\mathbb{R}^n} vu\Delta\chi\,\mathrm{d}y.$$

The last two terms are simplified by the following partial integration

$$\int_{\mathbb{R}^n} vu\Delta\chi\,\mathrm{d}y = -\int_{\mathbb{R}^n} \nabla\{vu\}\cdot\nabla\chi\,\mathrm{d}y = -\int_{\mathbb{R}^n} (u\nabla v + v\nabla u)\cdot\nabla\chi\,\mathrm{d}y.$$

Therefore, we have deduced the weak form of (3.12) :

$$\int_{\mathbb{R}^n} v\Delta\{\chi u\}\,\mathrm{d}y = \int_{\mathbb{R}^n} v\chi\Delta u\,\mathrm{d}y + \int_{\mathbb{R}^n} [v\nabla u - u\nabla v]\cdot\nabla\chi\,\mathrm{d}y. \tag{3.13}$$

The support of $\nabla\chi$ is restricted to the boundary, hence the second integral on the right-hand side encompasses all boundary integrals. Here, we only have first derivatives of the cutoff distribution χ. Thus, we do not get additional corner terms.

Distributional reciprocity

Equation (3.13) is almost the generalized Green's formula needed for the BEM. We have to integrate partially the integral on the left-hand side to get the *distributional reciprocity*, cf. Hörmander [Hör90], theorem 3.1.9. With $\Delta u = -f$, we get

$$\int_{\mathbb{R}^n} \chi u\Delta v\,\mathrm{d}y \quad = \quad -\int_{\mathbb{R}^n} v\chi f\,\mathrm{d}y - \int_{\mathbb{R}^n} [v\nabla u - u\nabla v]\cdot\nabla\chi\,\mathrm{d}y. \tag{3.14}$$

This procedure can be transferred to arbitrary differential operators. The last integral of (3.14) represents the boundary integrals, because the support of $\nabla\chi$ is restricted to $\partial\Omega$. The explicit equivalences are with $\nabla\chi = \delta(\psi)\nabla\psi$

$$\int_{\mathbb{R}^n} \delta(\psi)\nabla\psi\cdot\nabla u\,v\,\mathrm{d}y \quad = \quad -\int_\Gamma \nu\cdot\nabla u\,v\,\mathrm{d}\Gamma_y = -\int_\Gamma \frac{\partial u}{\partial\nu}v\,\mathrm{d}\Gamma_y; \tag{3.15}$$

$$\int_{\mathbb{R}^n} u\,\delta(\psi)\nabla\psi\cdot\nabla v\,\mathrm{d}y \quad = \quad -\int_\Gamma u\,\nu\cdot\nabla v\,\mathrm{d}\Gamma_y = -\int_\Gamma u\frac{\partial v}{\partial\nu}\,\mathrm{d}\Gamma_y, \tag{3.16}$$

which makes (3.14) comparable to (2.3). The main advantage of this representation is that the integrals extend formally over the entire $\mathbb{R}^n$. Therefore, the Fourier transformation can be applied to these integral equations.

3.3.2 Distributional Galerkin BIE

Distributional trial and test functions

For the definition of trial functions in analogy to (2.13), we have to use special cutoff distributions defined on the boundary panel. We start by defining a *cutoff distribution for a reference element* (Fig. 3.3)

$$\begin{aligned} \chi^0(x) &:= \mathrm{H}(x_1)\mathrm{H}(1-x_1)\,\delta(x_2), & x\in\mathbb{R}^2; \\ \chi^0(x) &:= \mathrm{H}(x_1)\mathrm{H}(1-x_1)\mathrm{H}(x_2)\mathrm{H}(1-x_2)\,\delta(x_3), & x\in\mathbb{R}^3. \end{aligned} \tag{3.17}$$

If a triangular mesh is preferred, we define

$$\chi^0(x) \;:=\; \mathrm{H}(x_1)\mathrm{H}(x_2)\mathrm{H}(1-x_1-x_2)\,\delta(x_3), \qquad x\in\mathbb{R}^3. \tag{3.18}$$

The trial functions are obtained by multiplying a $C^\infty(\mathbb{R}^n)$-function $p^0(x)$,

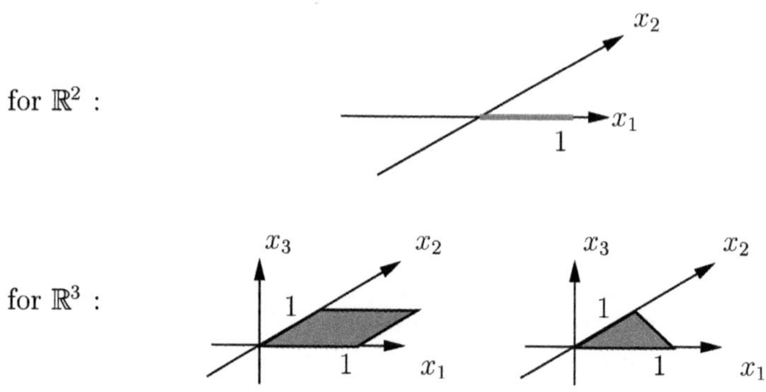

Figure 3.3: *Supports of the cutoff distributions for the reference elements*

e.g. a polynomial, with these cutoff distributions

$$\phi^0(x) \;:=\; \chi^0(x)\,p^0(x). \tag{3.19}$$

Arbitrary trial functions $\phi^i_{\mathrm{u,t}}(x)$ on arbitrary straight elements are obtained by translation and/or dilation operators

$$\mathcal{T}^i:\phi^0\to\phi^i_{\mathrm{u,t}}=\phi^0(x-b^i), \qquad \mathcal{D}^i:\phi^0\to\phi^i_{\mathrm{u,t}}=\phi^0(a^i x) \tag{3.20}$$

with the translation vector b^i and the dilation matrix a^i. Finally, the unknown and the known quantities on the boundaries are approximated by

$$\delta(\psi)u(x)\approx\sum_i^{N_\mathrm{u}}\mathrm{u}^i\phi^i_\mathrm{u}(x), \qquad \delta(\psi)\nabla\psi(x)\cdot\nabla u(x)\approx\sum_i^{N_\mathrm{t}}\mathrm{t}^i\phi^i_\mathrm{t}(x). \tag{3.21}$$

Distributional Galerkin BIE

In (3.14), v is now replaced by the fundamental solution $U(x-y)$ as in the derivation of (2.7). The additional weighting with ϕ_{t}^j of the form (3.20) leads to the distributional equivalent of the Galerkin BIE (2.14)

$$\begin{aligned}\int_{\mathbb{R}^n} \phi_{\mathrm{t}}^j(x)\chi(x)u(x)\,\mathrm{d}x &= \int_{\mathbb{R}^n} \phi_{\mathrm{t}}^j(x)\int_{\mathbb{R}^n} \chi(y)f(y)U(x-y)\,\mathrm{d}y\,\mathrm{d}x \\ &+ \int_{\mathbb{R}^n} \phi_{\mathrm{t}}^j(x)\int_{\mathbb{R}^n} \nabla\chi(y)\cdot\nabla u(y)U(x-y)\,\mathrm{d}y\,\mathrm{d}x \\ &- \int_{\mathbb{R}^n} \phi_{\mathrm{t}}^j(x)\int_{\mathbb{R}^n} u(y)\nabla\chi(y)\cdot\nabla U(x-y)\,\mathrm{d}y\,\mathrm{d}x.\end{aligned} \tag{3.22}$$

with $\nabla\chi = \delta(\psi)\nabla\psi$ for smooth boundaries. The term on the left-hand side was simplified due to[4]

$$\int_{\mathbb{R}^n} \phi_{\mathrm{t}}^j(x)\int_{\mathbb{R}^n} \chi(y)u(y)\,\delta(x-y)\,\mathrm{d}y\,\mathrm{d}x = \int_{\mathbb{R}^n} \phi_{\mathrm{t}}^j(x)\chi(x)u(x)\,\mathrm{d}x. \tag{3.23}$$

For smooth boundaries with $\chi = \mathrm{H}(\psi)$, we have $\delta(\psi)\mathrm{H}(\psi) = \delta(\psi)/2$ and therefore :

$$\int_{\mathbb{R}^n} \phi_{\mathrm{t}}^j(x)\mathrm{H}(\psi)u(x)\,\mathrm{d}x = \frac{1}{2}\int_{\mathbb{R}^n} \phi_{\mathrm{t}}^j(x)u(x)\,\mathrm{d}x. \tag{3.24}$$

System of Galerkin BIEs

To simplify the notation for the following chapters, the following abbreviations for the scalar product and the convolution are defined :

$$\text{scalar product:} \qquad \langle a,b\rangle = \int_{\mathbb{R}^n} a(x)\,b(x)\,\mathrm{d}x; \tag{3.25}$$

$$\text{convolution:} \qquad a*b = \int_{\mathbb{R}^n} a(y)\,b(x-y)\,\mathrm{d}y. \tag{3.26}$$

And by notating $u_\chi = u(x)\chi(x)$, $f_\chi = f(x)\chi(x)$ we get the extended Galerkin BIE for the temperature

$$\langle\phi_{\mathrm{t}}^j, u_\chi\rangle = \langle\phi_{\mathrm{t}}^j, f_\chi * U\rangle + \sum_i^{N_{\mathrm{t}}} \mathrm{t}^i\,\langle\phi_{\mathrm{t}}^j, \phi_{\mathrm{t}}^i * U\rangle - \sum_i^{N_{\mathrm{u}}} \mathrm{u}^i\,\langle\phi_{\mathrm{t}}^j, \phi_{\mathrm{u}}^i * \mathcal{A}_{\mathrm{t}}^i U\rangle \tag{3.27}$$

---

[4]The multiplication of the distributions ϕ_{t}^j and χ is defined in the context of non-linear distributions, cf. [Bia90, Boi96].

and for the flux :

$$-\langle\phi_{\rm u}^j, \mathcal{A}_{\rm t}^j u_\chi\rangle = -\langle\phi_{\rm u}^j, f_\chi * \mathcal{A}_{\rm t}^j U\rangle \tag{3.28}$$
$$- \sum_i^{N_{\rm t}} {\rm t}^i \langle\phi_{\rm u}^j, \phi_{\rm t}^i * \mathcal{A}_{\rm t}^j U\rangle + \sum_i^{N_{\rm u}} {\rm u}^i \langle\phi_{\rm u}^j, \phi_{\rm u}^i * \mathcal{A}_{\rm t}^j \mathcal{A}_{\rm t}^i U\rangle .$$

3.3.3 Regularized BIE

The fundamental solution U is weakly singular, its derivative $\mathcal{A}_{\rm t}^i U$ strongly, and $\mathcal{A}_{\rm t}^j \mathcal{A}_{\rm t}^i U$ hypersingular. Weakly singular kernels can be integrated easily, while particular regularization procedures are needed for the other cases that are often based on Cauchy's or Hadamard's Principal Values, cf. [Hör90]. Nevertheless, it can be shown that the BIE in total do not exhibit such singularities, i.e. that all non-integrable singularities cancel, compare [GKRR92, Bon95a].

Regularization of the temperature BIE

Due to $u(y)\delta(x-y) = u(x)\delta(x-y)$, cf. Hörmander [Hör90], example 3.1.2., we can write for the left-hand side of equation (3.22)

$$\int_{\mathbb{R}^n} \phi_{\rm t}^j(x) \int_{\mathbb{R}^n} \chi(y)u(y)\,\delta(x-y)\,{\rm d}y\,{\rm d}x \tag{3.29}$$
$$= \int_{\mathbb{R}^n} \phi_{\rm t}^j(x)u(x) \int_{\mathbb{R}^n} \chi(y)\delta(x-y)\,{\rm d}y\,{\rm d}x.$$

By applying an additional Green's formula (which is an equilibrium condition) we get with $\delta = -\Delta U = -{\rm div}\nabla U$

$$\int_{\mathbb{R}^n} \chi(y)\delta(x-y)\,{\rm d}y = -\int_{\mathbb{R}^n} \chi(y)\,{\rm div}\nabla U(x-y)\,{\rm d}y \tag{3.30}$$
$$= \int_{\mathbb{R}^n} \nabla\chi(y)\cdot\nabla U(x-y)\,{\rm d}y = \int_{\mathbb{R}^n} \delta(\psi)\nabla\psi\cdot\nabla U(x-y)\,{\rm d}y.$$

Hence, by inserting (3.30) in (3.22), the *regularized Galerkin BIE* for the temperature is obtained

$$\begin{aligned} 0 &= \langle\phi_{\rm t}^j(x), f_\chi(y) * U\rangle + \langle\phi_{\rm t}^j(x), (\nabla\chi\nabla u * U)\rangle \\ &- \langle\phi_{\rm t}^j(x), (u\nabla\chi * \nabla U)\rangle + \langle\phi_{\rm t}^j(x), u(\nabla\chi * \nabla U)\rangle . \end{aligned} \tag{3.31}$$

The introduction of the trial functions leads to the *regularized Galerkin BIE* for the temperature

$$
\begin{aligned}
0 &= \langle \phi_{\mathrm{t}}^j, f_\chi * U \rangle + \sum_i^{N_{\mathrm{t}}} \mathrm{t}^i \langle \phi_{\mathrm{t}}^i, \phi_{\mathrm{t}}^i * U \rangle \\
&- \sum_i^{N_{\mathrm{u}}} \mathrm{u}^i \langle \phi_{\mathrm{t}}^j, \phi_{\mathrm{u}}^i * \mathcal{A}_{\mathrm{t}}^i U \rangle + \sum_i^{N_{\mathrm{e}}} \mathrm{u}^j \langle \phi_{\mathrm{t}}^j p^j, \phi_{\mathrm{e}}^i * \mathcal{A}_{\mathrm{t}}^i U \rangle .
\end{aligned} \tag{3.32}
$$

with $N_{\mathrm{e}} = N_{\mathrm{t}} + N_{\mathrm{u}}$ as the total number of elements and ϕ_{e}^i as the constant trial function of the i-th element. We have simplified $\phi_{\mathrm{t}}^j u = \mathrm{u}^j \phi_{\mathrm{t}}^j p^j$ with p^j as the polynomial of the trial function ϕ_{t}^j. For $i \neq j$ all integrals are a priori regular, and for $i = j$ we obtained the regularized integrals

$$
\mathrm{u}^j \langle \phi_{\mathrm{t}}^j, \phi_{\mathrm{u}}^j * \mathcal{A}_{\mathrm{t}}^i U - p^j (\phi_{\mathrm{e}}^j * \mathcal{A}_{\mathrm{t}}^i U) \rangle . \tag{3.33}
$$

Regularization of the flux BIE

The left-hand side of the flux BIE (3.28) can be written as

$$
\int_{\mathbb{R}^n} \phi_{\mathrm{u}}^j \mathcal{A}_{\mathrm{t}}^j \{\chi u\} \,\mathrm{d}x = \int_{\mathbb{R}^n} \phi_{\mathrm{u}}^j \chi \mathcal{A}_{\mathrm{t}}^j u \,\mathrm{d}x + \int_{\mathbb{R}^n} \phi_{\mathrm{u}}^j u \mathcal{A}_{\mathrm{t}}^j \chi \,\mathrm{d}x. \tag{3.34}
$$

The last integral is divergent :

$$
\int_{\mathbb{R}^n} \phi_{\mathrm{u}}^j(x) u(x) \mathcal{A}_{\mathrm{t}}^j \chi(x) \,\mathrm{d}x = \int_{\mathbb{R}^n} \phi_{\mathrm{u}}^j(x) u(x) \delta(\psi) \mathcal{A}_{\mathrm{t}}^j \psi \,\mathrm{d}x. \tag{3.35}
$$

Example 3.3.1 Singularity of the free term

To visualize the singular character of (3.35), we consider the semicircle of Example 3.2.1 with a test function $\phi_{\mathrm{u}}^j = \delta(x_1)\mathrm{H}(x_2)\mathrm{H}(1 - x_2)$. *The singular integral is for the relevant part of the boundary with* $\psi = x_1, \nu = [1, 0]$

$$
\begin{aligned}
\int_{\mathbb{R}^n} \phi_{\mathrm{u}}^j(x) u(x) \delta(\psi) \mathcal{A}_{\mathrm{t}}^j \psi \,\mathrm{d}x &= \int_{\mathbb{R}^n} \delta(x_1)\mathrm{H}(x_2)\mathrm{H}(1 - x_2) u(x_1, x_2) \delta(x_1) \,\mathrm{d}x \\
&= \delta(0) \int_0^1 u(0, x_2) \,\mathrm{d}x_2.
\end{aligned}
$$

Therefore, the singular character is precisely $\delta(0)$.

This singularity is transferred to the right-hand side to regularize the hypersingular term. We obtain in analogy to (3.30) and (3.31) :

$$
\begin{aligned}
\int_{\mathbb{R}^n} \phi_{\mathrm{u}}^j(x) u(x) \mathcal{A}_{\mathrm{t}}^j \chi(x) \,\mathrm{d}x &= \int_{\mathbb{R}^n} \phi_{\mathrm{u}}^j(x) u(x) \mathcal{A}_{\mathrm{t}}^j \int_{\mathbb{R}^n} \chi(y)\, \delta(x - y) \,\mathrm{d}y \,\mathrm{d}x \\
&= \int_{\mathbb{R}^n} \phi_{\mathrm{u}}^j(x) u(x) \mathcal{A}_{\mathrm{t}}^j \int_{\mathbb{R}^n} \delta(\psi) \nabla \psi \cdot \nabla U(x - y) \,\mathrm{d}y \,\mathrm{d}x.
\end{aligned} \tag{3.36}
$$

Hence, the *regularized BIE for the flux* is

$$\begin{aligned} 0 &= \int_{\mathbb{R}^n} \phi_{\mathrm{u}}^j(x) \int_{\mathbb{R}^n} \chi(y) f(y)\, \mathcal{A}_{\mathrm{t}}^j U(x-y)\,\mathrm{d}y\,\mathrm{d}x \\ &- \int_{\mathbb{R}^n} \phi_{\mathrm{u}}^j(x) \int_{\mathbb{R}^n} \delta(\psi)[t(y)-t(x)]\, \mathcal{A}_{\mathrm{t}}^j U(x-y)\,\mathrm{d}y\,\mathrm{d}x \\ &+ \int_{\mathbb{R}^n} \phi_{\mathrm{u}}^j(x) \int_{\mathbb{R}^n} \delta(\psi)[u(y)-u(x)]\, \mathcal{A}_{\mathrm{t}}^j \mathcal{A}_{\mathrm{t}}^i U(x-y)\,\mathrm{d}y\,\mathrm{d}x. \end{aligned} \tag{3.37}$$

Chapter 4

Fourier BEM

Jean Baptiste Joseph Fourier

4.1 The Principal Idea

For several physical problems, we do not know the fundamental solution U required in (3.27) and (3.28). But (if the coefficients of the differential operator are constant), we can always derive the Fourier transform $\hat{U}(\hat{x})$ of $U(x)$. Therefore, Fourier–BEM as a more general numerical approach is derived in this chapter.

4.2 Fourier Transform of BIE

4.2.1 Transformed Galerkin BIE

The two fundamental theorems

The n-dimensional *Fourier transform* $\mathcal{F}(u) = \hat{u}$ of $u \in L^1(\mathbb{R}^n)$ is defined[1] as $(i = \sqrt{-1})$

$$\hat{u}(\hat{x}) = \int_{\mathbb{R}^n} u(x)\, \mathrm{e}^{-i<x,\hat{x}>}\, \mathrm{d}x, \qquad < x, \hat{x} > = \sum_k^n x_k \hat{x}_k. \tag{4.1}$$

[1]See Appendix A for a more general definition valid also for distributions.

It should be emphasized that the transformation has to be applied with respect to all coordinates $x_k \to \hat{x}_k$, $k = 1 \ldots n$; $(\hat{\cdot})$ characterizes a transformed object, $\overset{\mathcal{F}}{\leftrightarrow}$ denotes a pair of Fourier correspondence.

The basis of Fourier-BEM are two well known theorems of the Fourier transformation, the *theorem of Parseval*

$$\int_{\mathbb{R}^n} \phi(x)u(x)\,\mathrm{d}x = \frac{1}{(2\pi)^n}\int_{\mathbb{R}^n} \hat{\phi}(-\hat{x})\hat{u}(\hat{x})\,\mathrm{d}\hat{x}, \qquad x,\hat{x}\in\mathbb{R}^n, \tag{4.2}$$

and the *convolution theorem*, cf. [Bra86],

$$\int_{\mathbb{R}^n} \phi(y)u(x-y)\,\mathrm{d}y \quad \overset{\mathcal{F}}{\leftrightarrow} \quad \hat{\phi}(\hat{x})\,\hat{u}(\hat{x}). \tag{4.3}$$

Parseval's theorem (4.2) states the invariance of energy or work with respect to the n-dimensional Fourier transformation ($x, \hat{x} \in \mathbb{R}^n$). The convolution theorem (4.3) links a convolution in the original space to a simple multiplication in the transformed space. In the abbreviated notation introduced in (3.25), (3.26) we have for these two Fourier theorems :

$$\begin{aligned} \langle \phi(x), u(x)\rangle &= \frac{1}{(2\pi)^n}\left\langle \hat{\phi}(-\hat{x}), \hat{u}(\hat{x})\right\rangle; \qquad (4.4)\\ \phi(x) * u(x) &\overset{\mathcal{F}}{\leftrightarrow} \hat{\phi}(\hat{x})\,\hat{u}(\hat{x}). \end{aligned}$$

Transform of the convolutional part

The inner integrals in (3.27) are convolutions. By applying (4.3) they are converted by the Fourier transformation to simple multiplications

$$\begin{array}{lllll} f_\chi & * & U & \overset{\mathcal{F}}{\leftrightarrow} & \hat{f}_\chi(\hat{x}) \;\; \hat{U}(\hat{x}); \\ \phi_\mathrm{t}^i & * & U & \overset{\mathcal{F}}{\leftrightarrow} & \hat{\phi}_\mathrm{t}^i(\hat{x}) \;\; \hat{U}(\hat{x}); \\ \phi_\mathrm{u}^i & * & \mathcal{A}_\mathrm{t}^i U & \overset{\mathcal{F}}{\leftrightarrow} & \hat{\phi}_\mathrm{u}^i(\hat{x}) \;\; \hat{\mathcal{A}}_\mathrm{t}^i\hat{U}(\hat{x}). \end{array} \tag{4.5}$$

Transform of the scalar product

The outer integrals in (3.27) are scalar products (work), therefore we have to apply Parseval's theorem (4.2). Together with (4.5), we obtain as transform of (3.27) (the factor $(2\pi)^{-n}$ is cancelled)

$$\begin{aligned} \left\langle \hat{\phi}_\mathrm{t}^j(-\hat{x}), \hat{u}_\chi(\hat{x})\right\rangle &= \left\langle \hat{\phi}_\mathrm{t}^j(-\hat{x}), \hat{f}_\chi(\hat{x})\,\hat{U}(\hat{x})\right\rangle \qquad (4.6)\\ &+\sum_i^{N_\mathrm{t}} \mathrm{t}^i \left\langle \hat{\phi}_\mathrm{t}^j(-\hat{x}), \hat{\phi}_\mathrm{t}^i(\hat{x})\,\hat{U}(\hat{x})\right\rangle - \sum_i^{N_\mathrm{u}} \mathrm{u}^i \left\langle \hat{\phi}_\mathrm{t}^j(-\hat{x}), \hat{\phi}_\mathrm{u}^i(\hat{x})\,\hat{\mathcal{A}}_\mathrm{t}^i\hat{U}(\hat{x})\right\rangle. \end{aligned}$$

Transform of the free term

The free term, i.e. the left-hand side of (4.6), can be simplified due to

$$\phi_{\mathrm{t}}^{j} u_{\chi} = \phi_{\mathrm{t}}^{j}(x)\chi(x)u(x) = \sum_{i} \phi_{\mathrm{t}}^{j}(x)\mathrm{u}^{i} p_{\mathrm{u}}^{i}(x)\kappa(x),$$

where the sum is taken over all trial functions which have common support with ϕ_{t}^{j}. p_{u}^{i} are the polynomial of these trial functions. κ is the corner term discussed in Chapter 3.2. The polynomials p^{i} are transformed to Dirac-distributions, hence the evaluation of the scalar product on the left-hand side of (4.6) is trivial.

Example 4.2.1 Free term for smooth boundaries

As an example, we consider the two-dimensional half-space $x_1 \geq 0, x_2 \in \mathbb{R}$. For simplicity, a constant test function is chosen (although not conform), e.g.:

$$\phi_{\mathrm{t}}^{1} = \delta(x_1)\mathrm{H}(x_2)\mathrm{H}(1-x_2) \qquad \overset{\mathcal{F}}{\leftrightarrow} \qquad \hat{\phi}_{\mathrm{t}}^{1} = \frac{i}{\hat{x}_2}\left(\mathrm{e}^{-i\hat{x}_2} - 1\right).$$

u_{χ} is locally, i.e. at the support of ϕ_{t}^{1}, constant and equal to $\mathrm{u}^{1}\mathrm{H}(x_1) \overset{\mathcal{F}}{\leftrightarrow} \mathrm{u}^{1}(\pi\delta(\hat{x}_1) - i/\hat{x}_1)\, 2\pi\delta(\hat{x}_2)$. In the original space, we have then for the free term

$$\left\langle \phi_{\mathrm{t}}^{1}(x), \chi(x)\, u(x)\right\rangle = \mathrm{u}^{1} \int_{\mathbb{R}^2} \delta(x_1)\mathrm{H}(x_2)\mathrm{H}(1-x_2)\; \mathrm{H}(x_1)\, \mathrm{d}x_1\, \mathrm{d}x_2$$

$$= \mathrm{u}^{1}\left[\int_{-\infty}^{+\infty} \delta(x_1)\mathrm{H}(x_1)\, \mathrm{d}x_1\right]\left[\int_{0}^{1} 1\, \mathrm{d}x_2\right] = \mathrm{u}^{1}\int_{-\infty}^{+\infty} \delta(x_1)\mathrm{H}(x_1)\, \mathrm{d}x_1 = \frac{\mathrm{u}^{1}}{2}.$$

The last integral was evaluated by replacing the Dirac-distribution with a Dirac-convergent series, cf. (3.5). In the transformed space we obtain

$$\frac{1}{(2\pi)^2}\left\langle \hat{\phi}_{\mathrm{t}}^{1}(-\hat{x}), \hat{u}_{\chi}\right\rangle = \frac{\mathrm{u}^{1}}{2\pi}\int_{\mathbb{R}^2} \frac{-i}{\hat{x}_2}\left(\mathrm{e}^{+i\hat{x}_2} - 1\right)\left[\pi\delta(\hat{x}_1) - \frac{i}{\hat{x}_1}\right]\delta(\hat{x}_2)\, \mathrm{d}\hat{x}_1\, \mathrm{d}\hat{x}_2$$

$$= \frac{\mathrm{u}^{1}}{2\pi}\int_{-\infty}^{+\infty}\left(\pi\delta(\hat{x}_1) - \frac{i}{\hat{x}_1}\right)\mathrm{d}\hat{x}_1 = \frac{\mathrm{u}^{1}}{2} - \frac{\mathrm{u}^{1}}{2\pi}\int_{-\infty}^{+\infty}\frac{i}{\hat{x}_1}\,\mathrm{d}\hat{x}_1 = \frac{\mathrm{u}^{1}}{2}.$$

Here, we have regularized the last integral, i.e. we have taken a Cauchy's Principal Value at infinity: the singularity at $\hat{x}_1 = +\infty$ cancels with that at $\hat{x}_1 = -\infty$. The kernel $1/\hat{x}_1$ is odd, hence the integral over the total real line is zero. We have evaluated the free term as in the original space for smooth boundaries.

Transform of the BIE for temperature and flux

The *transform of the Galerkin BIE* given in (3.27) and (3.28) is

$$\begin{aligned}\left\langle \hat{\phi}_{\mathrm{t}}^{j}(-\hat{x}), \hat{u}_{\chi} \right\rangle &= \left\langle \hat{\phi}_{\mathrm{t}}^{j}(-\hat{x}), \hat{f}_{\chi}(\hat{x})\, \hat{U}(\hat{x}) \right\rangle + \sum_{i}^{N_{\mathrm{t}}} \mathrm{t}^{i} \left\langle \hat{\phi}_{\mathrm{t}}^{j}(-\hat{x}), \hat{\phi}_{\mathrm{t}}^{i}(\hat{x})\, \hat{U}(\hat{x}) \right\rangle \\ &- \sum_{i}^{N_{\mathrm{u}}} \mathrm{u}^{i} \left\langle \hat{\phi}_{\mathrm{t}}^{j}(-\hat{x}), \hat{\phi}_{\mathrm{u}}^{i}(\hat{x}) \hat{\mathcal{A}}_{\mathrm{t}}^{i} \hat{U}(\hat{x}) \right\rangle ; \qquad (4.7) \\ -\left\langle \hat{\phi}_{\mathrm{u}}^{j}(-\hat{x}), \hat{\mathcal{A}}_{\mathrm{t}}^{j} \hat{u}_{\chi} \right\rangle &= -\left\langle \hat{\phi}_{\mathrm{u}}^{j}(-\hat{x}), \hat{f}_{\chi}(\hat{x})\, \hat{\mathcal{A}}_{\mathrm{t}}^{j} \hat{U}(\hat{x}) \right\rangle \\ &- \sum_{i}^{N_{\mathrm{t}}} \mathrm{t}^{i} \left\langle \hat{\phi}_{\mathrm{u}}^{j}(-\hat{x}), \hat{\phi}_{\mathrm{t}}^{i}(\hat{x})\, \hat{\mathcal{A}}_{\mathrm{t}}^{j} \hat{U}(\hat{x}) \right\rangle \\ &+ \sum_{i}^{N_{\mathrm{u}}} \mathrm{u}^{i} \left\langle \hat{\phi}_{\mathrm{u}}^{j}(-\hat{x}), \hat{\phi}_{\mathrm{u}}^{i}(\hat{x})\, \hat{\mathcal{A}}_{\mathrm{t}}^{j} \hat{\mathcal{A}}_{\mathrm{t}}^{i} \hat{U}(\hat{x}) \right\rangle . \qquad (4.8)\end{aligned}$$

4.3 Transformed fundamental solutions

4.3.1 Fundamental solutions

Transform of the fundamental solution

The Fourier transformation of the differential equation converts the differential operator $\mathcal{P}(\partial)$ to an algebraic expression $\hat{\mathcal{P}}(\hat{x})$

$$\Delta u(x) = -f(x) \quad \overset{\mathcal{F}}{\leftrightarrow} \quad -|\hat{x}|^2 \hat{u}(\hat{x}) = -\hat{f}(\hat{x}), \qquad (4.9)$$

with the symbol $\hat{\mathcal{P}} = -|\hat{x}|^2 = -\sum_k^n \hat{x}_k^2$. The Fourier-fundamental solution $\hat{U}$ is the response to a single unit force $f(x) = \delta(x) \overset{\mathcal{F}}{\leftrightarrow} \hat{f}(\hat{x}) = 1$. Therefore, we have to solve

$$\Delta U(x) = -\delta(x) \quad \overset{\mathcal{F}}{\leftrightarrow} \quad -|\hat{x}|^2 \hat{U}(\hat{x}) = -1, \qquad (4.10)$$

which is achieved in the transformed space by the inversion of $\hat{\mathcal{P}}$

$$\hat{U}(\hat{x}) = \frac{1}{|\hat{x}|^2}. \qquad (4.11)$$

This procedure can be applied to all linear differential operators with constant coefficients. The Fourier-fundamental solution is always known and often has a simple structure. $\hat{U}$ is singular in the point $|\hat{x}| = 0$ and hence not uniquely determined at the origin, cf. the mathematical discussion in [DP01]. The zeroes of the denominator correspond to additional homogeneous solutions of the differential equation

$$\Delta u(x) = 0 \quad \overset{\mathcal{F}}{\leftrightarrow} \quad -|\hat{x}|^2 \hat{u}(\hat{x}) = 0. \tag{4.12}$$

Transform of the derivatives of the fundamental solution

Due to our restriction to straight elements, the normal vector ν^i is locally independent of x. Hence we get as transform of the fundamental flux and of the hyper singular term

$$\begin{aligned} \mathcal{A}_t^i U = -\nu^i \cdot \nabla U \quad &\overset{\mathcal{F}}{\leftrightarrow} \quad \hat{\mathcal{A}}_t^i \hat{U} = -\nu^i \cdot i\hat{x}\hat{U}; \\ \mathcal{A}_t^j \mathcal{A}_t^i U = \nu^j \cdot \nabla(\nu^i \cdot \nabla U) \quad &\overset{\mathcal{F}}{\leftrightarrow} \quad \hat{\mathcal{A}}_t^j \hat{\mathcal{A}}_t^i \hat{U} = \nu^j \cdot i\hat{x}(\nu^i \cdot i\hat{x})\hat{U}. \end{aligned} \tag{4.13}$$

4.3.2 Green's functions

Green's function for the half-space

It is often advantageous to establish a BEM based on Green's functions $V(x)$, i.e. on special fundamental solutions which already fulfill some of the required boundary conditions. Discretization is then only needed in the other parts of the boundary; fewer degrees of freedom are required, e.g. [Ric96].

For half-space problems, Green's functions in the original space can be evaluated by the method of images if the fundamental solution is known, see e.g. [Bon95a]. Because the proposed Fourier-BEM is especially of interest for cases where we do not know this fundamental solution we need an alternative approach for the generation of Green's functions which should be totally formulated in the transformed space. This method was developed in Duddeck [Dud97b].

The condition that V should vanish for $x_1 < 0$ is equivalent to a condition derived from the theorem of Paley-Wiener (cf. [Hör90], theorem 7.3.1.). The transform of a quantity which vanishes in the half-space $x_1 < 0$ has no singularities in the lower complex half-plane $\Im(\hat{x}_1) \leq 0$ of the transformed coordinates, cf. [Dud97b, Wol79]. Hence we are looking for additional terms which eliminate the singularities of the fundamental solution $\hat{U}$ for $\Im(\hat{x}_1) \leq 0$.

Example 4.3.1 Two-dimensional half-space

We consider as an example the two-dimensional half-space $x_1 \geq 0, x_2 \in \mathbb{R}$

with a flux-free boundary $\partial\Omega$. The temperature $\psi(0,x_2)$ at $x_1 = 0$ can be regarded as external sources described by

$$f(x) = \psi(0,x_2)\delta'(x_1) \qquad \overset{\mathcal{F}}{\leftrightarrow} \qquad \hat{f}(\hat{x}) = i\hat{x}_1\hat{\psi}(0,\hat{x}_2).$$

The corresponding temperature $f * U \overset{\mathcal{F}}{\leftrightarrow} \hat{f}\hat{U}$ in transformed coordinates subtracted from the fundamental solution $\hat{U}$ leads to Green's function for the half-space $\hat{V}$ with still unknown ψ :

$$\hat{V} = \hat{U}\left[1 - i\hat{x}_1\hat{\psi}(0,\hat{x}_2)\right] = \frac{1 - i\hat{x}_1\hat{\psi}(0,\hat{x}_2)}{\hat{x}_1^2 + \hat{x}_2^2}.$$

The fundamental solution for the infinite space $\hat{U} = |\hat{x}|^{-2}$ is singular at $\hat{x}_1 = \pm i\hat{x}_2$. These singularities should cancel for $\hat{x}_1 = -i|\hat{x}_2|$. Hence we have to annul the residue

$$\mathrm{Res}[\hat{V}, \hat{x}_1 = -i|\hat{x}_2|] = \frac{1}{\hat{x}_2}\left[1 - i|\hat{x}_2|\hat{\psi}(0,\hat{x}_2)\right] \overset{!}{=} 0.$$

This leads to the unknown temperature at the boundary $\hat{\psi}(0,\hat{x}_2) = |\hat{x}_2|^{-1}$ and finally to the transform of the Green's function for the half-space

$$\hat{V}(\hat{x}) = \frac{1}{|\hat{x}|^2}\left[1 - \frac{i\hat{x}_1}{|\hat{x}_2|}\right].$$

This Fourier analogue of the method of images can be applied to all differential problems with constant coefficients in arbitrary dimensions and can as well be adapted to determine causal time-dependent fundamental solutions in the transformed space, cf. [Dud97b].

Green's function for the ball

As second example, the Green's function $\hat{V}$ for an n-dimensional ball is determined where the temperature at the boundary is assumed to be zero. $\hat{V}$ is not only useful for circular problems, but for all bounded problems, if the domain Ω is entirely enclosed in the ball $B(x)$, Fig. 4.1. It can be deduced from the Paley-Wiener theorem that $\hat{V}(\hat{x})$ do not have any singularities at all. Hence, no singular integrals have to be computed. The approach can be applied to all isotropic problems.

Example 4.3.2 Two-dimensional disc

The constant flux at the boundary of a two-dimensional ball with radius r_0 as external source is, cf. [Dud97b],

$$f(x) = 2\,r_0\psi(r_0)\delta(r_0^2 - x_1^2 - x_2^2) \qquad \overset{\mathcal{F}}{\leftrightarrow} \qquad \hat{f}(\hat{x}) = 2\pi\,r_0\psi(r_0)\mathrm{J}_0(r_0|\hat{x}|),$$

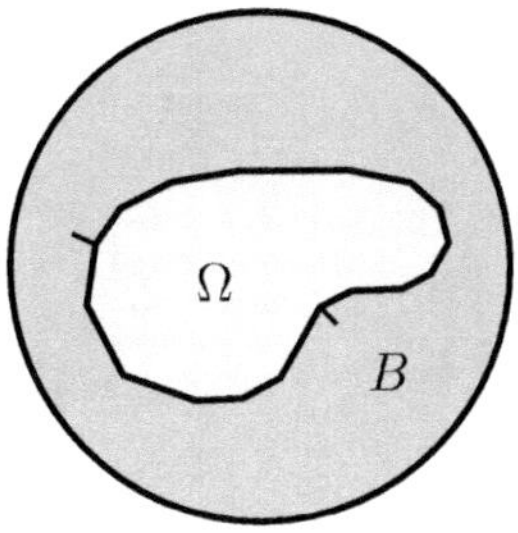

Figure 4.1: *The ball $B \subset \mathbb{R}^n$ which contains the total domain, i.e.: $\Omega \Subset B$.*

with J_0 *as Bessel's function of order zero. The Green's function is then :*

$$\hat{V} = \hat{U}\left[1 - 2\pi\, r_0 \psi(r_0) \mathrm{J}_0(r_0|\hat{x}|)\right] = \frac{1 - 2\pi\, r_0 \psi(r_0) \mathrm{J}_0(r_0|\hat{x}|)}{\hat{x}_1^2 + \hat{x}_2^2}.$$

The singularities, i.e. the residues, have to vanish (Paley-Wiener)

$$\mathrm{Res}\left[\mathrm{Res}\left[\frac{1 - 2\pi\, r_0 \psi(r_0) \mathrm{J}_0(r_0|\hat{x}|)}{\hat{x}_1^2 + \hat{x}_2^2}, \hat{x}_1 = i\hat{x}_2\right], \hat{x}_2 = 0\right] = -\frac{i}{2} + i\pi r_0 \psi(r_0) \stackrel{!}{=} 0.$$

Hence, we get for the flux $\psi(r_0) = (2\pi r_0)^{-1}$*, and finally for Green's function*

$$\hat{V}(\hat{x}) = \frac{1 - \mathrm{J}_0(r_0|\hat{x}|)}{\hat{x}_1^2 + \hat{x}_2^2}.$$

The non-singular character of $\hat{V}$ *and its dependency on the radius of the ball* B *is shown in Fig. 4.2. The larger the radius, the smaller is the extension of the Green's function. The residual approach shown here is a generalization of an equilibrium condition. In the simple case discussed here, the flux* $\psi(r_0)$ *can as well be determined by enforcing the static equilibrium between the single unit source at the origin and the circular flux at the boundary*

$$\int_{\mathbb{R}^2} \left[\delta(x) - \psi(r_0)\, \delta(r_0 - |x|)\right] \mathrm{d}x = 1 - 2\pi r_0 \psi(r_0) \stackrel{!}{=} 0.$$

4.4 Transformed Trial and Test Functions

4.4.1 Transform of the cutoff distributions

The transform of the cutoff distribution χ^0 is for the reference element in $\mathbb{R}^2$

$$\chi^0(x) = \mathrm{H}(x_1)\mathrm{H}(1 - x_1)\delta(x_2) \qquad \stackrel{\mathcal{F}}{\leftrightarrow} \qquad \hat{\chi}^0(\hat{x}) = \frac{i}{\hat{x}_1}\left(\mathrm{e}^{-i\hat{x}_1} - 1\right). \quad (4.14)$$

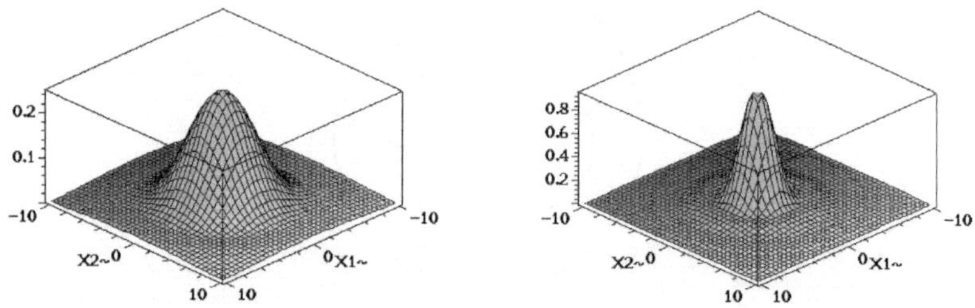

Figure 4.2: Green's function for the disc B with the radius $r_0 = 1$ (left) and $r_0 = 2$ (right) in the Fourier space.

For the quadratic boundary element in $\mathbb{R}^3$ we get

$$\begin{aligned} \chi^0(x) &= \mathrm{H}(x_1)\mathrm{H}(1-x_1)\mathrm{H}(x_2)\mathrm{H}(1-x_2)\delta(x_3) \\ \overset{\mathcal{F}}{\leftrightarrow} \quad \hat{\chi}^0(\hat{x}) &= \frac{i}{\hat{x}_1}\left(\mathrm{e}^{-i\hat{x}_1}-1\right)\frac{i}{\hat{x}_2}\left(\mathrm{e}^{-i\hat{x}_2}-1\right), \end{aligned} \tag{4.15}$$

and for the triangular boundary element in $\mathbb{R}^3$

$$\begin{aligned} \chi^0(x) &= \mathrm{H}(x_1)\mathrm{H}(x_2)\mathrm{H}(1-x_1-x_2)\delta(x_3) \\ \overset{\mathcal{F}}{\leftrightarrow} \quad \hat{\chi}^0(\hat{x}) &= \frac{1}{\hat{x}_1-\hat{x}_2}\left[\frac{1}{\hat{x}_1}\left(1-\mathrm{e}^{-i\hat{x}_1}\right)-\frac{1}{\hat{x}_2}\left(1-\mathrm{e}^{-i\hat{x}_2}\right)\right]. \end{aligned} \tag{4.16}$$

4.4.2 Transform of the trial functions

Elements of arbitrary polynomial degree are constructed via a multiplication by $p^0(x)$ in the original space or an analytical convolution in the transformed space

$$\phi^0(x) = \chi^0(x)p^0(x) \quad \overset{\mathcal{F}}{\leftrightarrow} \quad \hat{\phi}^0(\hat{x}) = \frac{1}{(2\pi)^n}\hat{\chi}^0(\hat{x}) * \hat{p}^0(\hat{x}). \tag{4.17}$$

All polynomials $p^0(x)$ are transformed to Dirac distributions

$$p^0(x) = \sum_k \sum_l c_{kl} x_k^l \quad \overset{\mathcal{F}}{\leftrightarrow} \quad \hat{p}^0(\hat{x}) = (2\pi)^n \sum_k \sum_l c_{kl} i^l \delta^{(l)}(\hat{x}_k).$$

Hence, the evaluation of the convolutions in (4.17) is trivial

$$\frac{1}{(2\pi)^n}\hat{p}^0 * \hat{\chi}^0 = \sum_k \sum_l c_{kl} i^l \partial_k^l \hat{\chi}^0$$

Table 4.1 shows the trial functions for some two-dimensional line elements. Three-dimensional elements are constructed analoguously.

constant:	$\phi^0 = \chi^0$	$\overset{\mathcal{F}}{\leftrightarrow}$	$\hat{\phi}^0 = \hat{\chi}^0$
linear:	$\phi^1 = (1 - x_1)\chi^0$	$\overset{\mathcal{F}}{\leftrightarrow}$	$\hat{\phi}^1 = (1 - i\partial_1)\hat{\chi}^0$
	$\phi^2 = x_1\chi^0$	$\overset{\mathcal{F}}{\leftrightarrow}$	$\hat{\phi}^2 = i\partial_1\hat{\chi}^0$
parabolic:	$\phi^1 = (2x_1^2 - 3x_1 + 1)\chi^0$	$\overset{\mathcal{F}}{\leftrightarrow}$	$\hat{\phi}^1 = (-2\partial_1^2 - 3i\partial_1 + 1)\hat{\chi}^0$
	$\phi^2 = (4x_1^2 - 4x_1)\chi^0$	$\overset{\mathcal{F}}{\leftrightarrow}$	$\hat{\phi}^2 = (-4\partial_1^2 - 4i\partial)\hat{\chi}^0$
	$\phi^3 = (2x_1^2 - x_1)\chi^0$	$\overset{\mathcal{F}}{\leftrightarrow}$	$\hat{\phi}^3 = (-2\partial_1^2 - i\partial_1)\hat{\chi}^0$

Table 4.1: *Trial and test functions of the* $\mathbb{R}^2$

General transformed straight elements are obtained as shown in (3.20) by the transformed dilation and translation operators

$$\begin{aligned} \mathcal{T}^i : \phi^0 \to \phi^i = \phi^0(x - b^i) \;&\overset{\mathcal{F}}{\leftrightarrow}\; \hat{\mathcal{T}}^i : \hat{\phi}^0 \to \hat{\phi}^i = \hat{\phi}^0(\hat{x})e^{-i<b^i,\hat{x}>} \\ \mathcal{D}^i : \phi^0 \to \phi^i = \phi^0(a^i x) \;&\overset{\mathcal{F}}{\leftrightarrow}\; \hat{\mathcal{D}}^i : \hat{\phi}^0 \to \hat{\phi}^i = |\det a^i|^{-1}\hat{\phi}^0([a^i]^{-T}\hat{x}), \end{aligned} \tag{4.18}$$

with $[a^i]^{-T}$ as the transposed inverse matrix of a^i. For straight elements and for arbitrary polynomial trial functions $p^0(x)$, the transformed expressions are analytically known for $\mathbb{R}^2$ and $\mathbb{R}^3$.

4.5 Construction of the BEM matrices

Algebraic system of equations

The discretized Fourier BIE (4.7) and (4.8) lead to an algebraic system of j equations *identical* to that obtained in the original space (2.26) :

$$\begin{aligned} \text{from (4.7)} : \quad \sum_i K_{\mathrm{u}}^{ji}\mathrm{u}^i &= F_{\mathrm{u}}^j + \sum_i H_{\mathrm{u}}^{ji}\mathrm{t}^i - \sum_i G_{\mathrm{u}}^{ji}\mathrm{u}^i; \\ \text{from (4.8)} : \quad \sum_i K_{\mathrm{t}}^{ji}\mathrm{t}^i &= F_{\mathrm{t}}^j + \sum_i H_{\mathrm{t}}^{ji}\mathrm{t}^i - \sum_i G_{\mathrm{t}}^{ji}\mathrm{u}^i. \end{aligned} \tag{4.19}$$

But now, the matrices are computed in the transformed space. We have :

$$
\begin{aligned}
F_{\mathrm{u}}^{j} &= \frac{1}{(2\pi)^n}\left\langle \hat{\phi}_{\mathrm{t}}^{j}(-\hat{x}), \hat{f}(\hat{x})\,\hat{U}(\hat{x})\right\rangle; \\
H_{\mathrm{u}}^{ji} &= \frac{1}{(2\pi)^n}\left\langle \hat{\phi}_{\mathrm{t}}^{j}(-\hat{x}), \hat{\phi}_{\mathrm{t}}^{i}(\hat{x})\,\hat{U}(\hat{x})\right\rangle; \\
G_{\mathrm{u}}^{ji} &= \frac{1}{(2\pi)^n}\left\langle \hat{\phi}_{\mathrm{t}}^{j}(-\hat{x}), \hat{\phi}_{\mathrm{u}}^{i}(\hat{x})\,\hat{A}_{\mathrm{t}}^{i}\hat{U}(\hat{x})\right\rangle; \\
K_{\mathrm{u}}^{ji} &:= \frac{1}{(2\pi)^n}\left\langle \hat{\phi}_{\mathrm{t}}^{j}(-\hat{x}), \hat{p}_{\mathrm{u}}^{i}(\hat{x})\right\rangle; \\
F_{\mathrm{t}}^{j} &= \frac{1}{(2\pi)^n}\left\langle \hat{\phi}_{\mathrm{u}}^{j}(-\hat{x}), \hat{f}(\hat{x})\,\hat{A}_{\mathrm{t}}^{j}\hat{U}(\hat{x})\right\rangle; \\
H_{\mathrm{t}}^{ji} &:= \frac{1}{(2\pi)^n}\left\langle \hat{\phi}_{\mathrm{u}}^{j}(\hat{x}), \hat{\phi}_{\mathrm{t}}^{i}(\hat{x})\,\hat{A}_{\mathrm{t}}^{j}\hat{U}(\hat{x})\right\rangle; \\
G_{\mathrm{t}}^{ji} &:= \frac{1}{(2\pi)^n}\left\langle \hat{\phi}_{\mathrm{u}}^{j}(\hat{x}), \hat{\phi}_{\mathrm{u}}^{i}(\hat{x})\,\hat{A}_{\mathrm{t}}^{j}\hat{A}_{\mathrm{t}}^{i}\hat{U}(\hat{x})\right\rangle; \\
K_{\mathrm{t}}^{ji} &:= \frac{1}{(2\pi)^n}\left\langle \hat{\phi}_{\mathrm{u}}^{j}(-\hat{x}), \hat{p}_{\mathrm{t}}^{i}\kappa^{i} + \hat{A}_{\mathrm{t}}^{j}\hat{p}_{\mathrm{u}}^{i}\kappa^{i}\right\rangle.
\end{aligned}
\tag{4.20}
$$

Due to the equivalence of the work terms in the original space and the transformed space which is stated by Parseval's theorem (4.2), all the vectors and matrices of (4.19) have the same values as would be obtained by a traditional BEM approach. Therefore, the further algorithm of the BEM algorithm can be taken without any modification from the standard BEM. This is not discussed here, see [Bon99].

Seperation of known and unknown boundary values

The final BEM algorithm is obtained by rearranging the matrices obtained in (4.19). The parts related to the known boundary values have to be shifted to the right-hand side, i.e. to the force vector, of the equation. The reordered system can be written in the following form

$$\sum_i A^{ji} X^i = Y^j. \tag{4.21}$$

with the symmetric and fully populated matrix A and the vector of unknown degrees of freedom X.

Chapter 5

Heat Conduction

Siméon Denis Poisson

5.1 Isotropic Case

5.1.1 The Dirichlet problem

We consider first a plane domain Ω which is heated by stationary interior sources f. The temperature at the boundaries is kept to zero. Mathematically, this leads to the Dirichlet problem of the Poisson equation. Due to the lack of singularities and its general simplicity, this Dirichlet problem is best suited to demonstrating the equivalence between traditional BEM and Fourier BEM. Without loss of generality, the isotropic conductivity K is set to one.

The Fourier-Galerkin boundary integral equations

As a first and benchmark example, the Dirichlet problem for the Poisson equation

$$\begin{aligned} \Delta u(x) &= -f(x), \qquad x \in \Omega; \\ u(x) &= u_\Gamma = 0, \qquad x \in \Gamma, \end{aligned} \tag{5.1}$$

is solved in a quadratic two-dimensional domain $\Omega = [0,1] \times [0,1]$. At the boundaries the temperature u is kept to zero. The interior is subjected to stationary heat sources f. The boundary $\partial\Omega$ is divided into eight elements, two for each side (Fig.5.1). We prefer this coarse mesh for the sake of clarity of the validation. The fundamental solution and its transform for the Laplacian Δ are

$$U = \frac{1}{2\pi} \ln \sqrt{x_1^2 + x_2^2} \qquad \overset{\mathcal{F}}{\leftrightarrow} \qquad \hat{U} = -\frac{1}{\hat{x}_1^2 + \hat{x}_2^2} \tag{5.2}$$

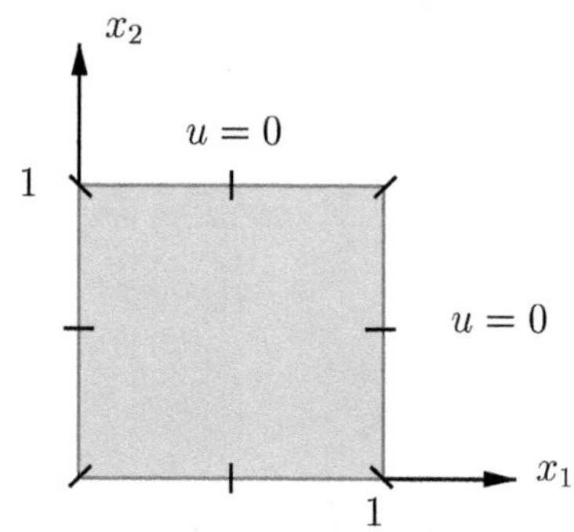

Figure 5.1: Quadratic domain with eight boundary elements

Taking into account that the temperature is vanishing at the boundaries, the general system of BIE can be reduced to

$$
\begin{aligned}
0 &= \langle \phi_{\mathrm{t}}^{j}, f_{\chi} * U \rangle + \sum_{i}^{N_{\mathrm{t}}} \mathrm{t}^{i} \langle \phi_{\mathrm{t}}^{j}, \phi_{\mathrm{t}}^{i} * U \rangle \qquad (5.3) \\
\overset{\mathcal{F}}{\leftrightarrow} \quad 0 &= \left\langle \hat{\phi}_{\mathrm{t}}^{j}(-\hat{x}), \hat{f}_{\chi} \hat{U} \right\rangle + \sum_{i}^{N_{\mathrm{t}}} \mathrm{t}^{i} \left\langle \hat{\phi}_{\mathrm{t}}^{j}(-\hat{x}), \hat{\phi}_{\mathrm{t}}^{i} \hat{U} \right\rangle .
\end{aligned}
$$

A uniform heat source f is assumed over Ω (Fig.5.2, right) :

$$
\begin{aligned}
f_{\chi}(x) &= f_0 \,\mathrm{H}(x_1)\mathrm{H}\,(1 - x_1)\,\mathrm{H}(x_2)\mathrm{H}\,(1 - x_2) \qquad (5.4) \\
\overset{\mathcal{F}}{\leftrightarrow} \quad \hat{f}_{\chi}(\hat{x}) &= -f_0 \,\frac{(\mathrm{e}^{-i\hat{x}_1} - 1)(\mathrm{e}^{-i\hat{x}_2} - 1)}{\hat{x}_1 \hat{x}_2} .
\end{aligned}
$$

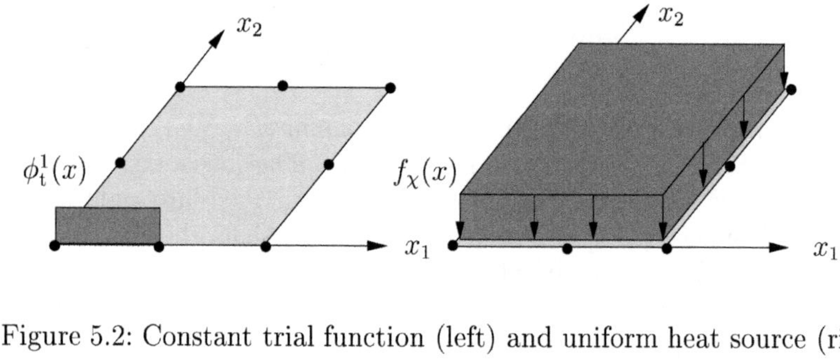

Figure 5.2: Constant trial function (left) and uniform heat source (right)

Equivalence test of BIE in Fourier space

The constant trial and test functions for the flux t, Fig.5.2, left,

$$
\begin{array}{lllll}
\phi_t^1 &=& \mathrm{H}(x_1)\mathrm{H}(1-2x_1)\,\delta(x_2) & \overset{\mathcal{F}}{\leftrightarrow} \ \hat\phi_t^1 &=& i\hat x_1^{-1}(\mathrm{e}^{-i\hat x_1/2}-1),\\
\phi_t^2 &=& \mathrm{H}(2x_1-1)\mathrm{H}(1-x_1)\delta(x_2) & \overset{\mathcal{F}}{\leftrightarrow} \ \hat\phi_t^2 &=& i\hat x_1^{-1}(\mathrm{e}^{-i\hat x_1}-\mathrm{e}^{-i\hat x_1/2}),\\
\phi_t^3 &=& \mathrm{H}(x_2)\mathrm{H}(1-2x_2)\delta(x_1-1) & \overset{\mathcal{F}}{\leftrightarrow} \ \hat\phi_t^3 &=& i\hat x_2^{-1}(\mathrm{e}^{-i\hat x_2/2}-1)\mathrm{e}^{-i\hat x_1},\\
\phi_t^4 &=& \mathrm{H}(2x_2-1)\mathrm{H}(1-x_2)\delta(x_1-1) & \overset{\mathcal{F}}{\leftrightarrow} \ \hat\phi_t^4 &=& i\hat x_2^{-1}(\mathrm{e}^{-i\hat x_2}-\mathrm{e}^{-i\hat x_2/2})\mathrm{e}^{-i\hat x_1},\\
\phi_t^5 &=& \mathrm{H}(x_1)\mathrm{H}(1-2x_1)\delta(x_2-1) & \overset{\mathcal{F}}{\leftrightarrow} \ \hat\phi_t^5 &=& i\hat x_1^{-1}(\mathrm{e}^{-i\hat x_1/2}-1)\mathrm{e}^{-i\hat x_2},\\
\phi_t^6 &=& \mathrm{H}(2x_1-1)\mathrm{H}(1-x_1)\delta(x_2-1) & \overset{\mathcal{F}}{\leftrightarrow} \ \hat\phi_t^6 &=& i\hat x_1^{-1}(\mathrm{e}^{-i\hat x_1}-\mathrm{e}^{-i\hat x_1/2})\mathrm{e}^{-i\hat x_2},\\
\phi_t^7 &=& \mathrm{H}(x_2)\mathrm{H}(1-2x_2)\delta(x_1) & \overset{\mathcal{F}}{\leftrightarrow} \ \hat\phi_t^7 &=& i\hat x_2^{-1}(\mathrm{e}^{-i\hat x_2/2}-1),\\
\phi_t^8 &=& \mathrm{H}(2x_2-1)\mathrm{H}(1-x_2)\delta(x_1) & \overset{\mathcal{F}}{\leftrightarrow} \ \hat\phi_t^8 &=& i\hat x_2^{-1}(\mathrm{e}^{-i\hat x_2}-\mathrm{e}^{-i\hat x_2/2}).
\end{array}
\tag{5.5}
$$

The integration of the first matrix entry in the original space is

$$
\begin{aligned}
H^{11} &= \int_0^{1/2}\!\!\int_0^{1/2} U(x_1-y_1)\,\mathrm{d}y_1\,\mathrm{d}x_1 = \frac{1}{2\pi}\int_0^{1/2}\!\!\int_0^{1/2} \ln\sqrt{(x_1-y_1)^2}\,\mathrm{d}y_1\,\mathrm{d}x_1\\
&= \frac{1}{8\pi}\int_0^{1/2}\Big[\ln|x_1-1/2|^2-2-4x_1(\ln|x_1-1/2|-\ln|x_1|)\Big]\,\mathrm{d}x_1\\
&= -\frac{1}{16\,\pi}(2\ln 2+3) = -0.08726.
\end{aligned}
\tag{5.6}
$$

The corresponding integration in Fourier space leads to the same value

$$
\begin{aligned}
H^{11} &= \frac{1}{(2\pi)^2}\int_{\mathbb{R}^2}\hat\phi_t^1(-\hat x)\hat\phi_t^1(\hat x)\,\hat U(\hat x)\,\mathrm{d}\hat x \qquad (5.7)\\
&= \frac{1}{(2\pi)^2}\int_{\mathbb{R}^2}\frac{[i\,(\mathrm{e}^{+i\hat x_1/2}-1)][i\,(\mathrm{e}^{-i\hat x_1/2}-1)]}{-\hat x_1^2(-\hat x_1^2-\hat x_2^2)}\,\mathrm{d}\hat x_1\,\mathrm{d}\hat x_2\\
&= \frac{1}{2\pi}\int_{\mathbb{R}^1}\frac{\operatorname{sgn}\hat x_1}{\hat x_1^3}\,[\cos(\hat x_1/2)-1]\,\mathrm{d}\hat x_1 = -\frac{1}{16\,\pi}(2\ln 2+3).
\end{aligned}
$$

The benchmark example

The total matrix H is obtained analytically either in the original or in Fourier space as :

$$
H = \begin{bmatrix}
-.08726 & -.03210 & -.00934 & .00234 & .00079 & .00484 & -.04222 & .00934\\
 & -.08726 & -.04222 & -.00934 & .00484 & .00079 & -.00934 & .00234\\
 & & -.08726 & -.03210 & .00234 & -.00934 & .00079 & .00484\\
 & & & -.08726 & -.00934 & -.04222 & .00484 & .00079\\
 & & & & -.08726 & -.03210 & -.00934 & -.04222\\
 & & \text{sym.} & & & -.08726 & .00234 & -.00934\\
 & & & & & & -.08726 & -.03210\\
 & & & & & & & -.08726
\end{bmatrix}
$$

The in tegration of volume sources (Fig.5.2, right) leads for the first component of F^k to ($f_0 = 1$)

$$\begin{aligned} F^1 &= \frac{1}{(2\pi)^2}\int_{\mathbb{R}^2} \hat{\phi}^1_t(-\hat{x})\hat{f}_\chi(\hat{x})\hat{U}(\hat{x})\,\mathrm{d}\hat{x} \qquad (5.8)\\ &= \frac{f_0}{(2\pi)^2}\int_{\mathbb{R}^2}\left[\frac{i}{\hat{x}_1}\left(\mathrm{e}^{+i\hat{x}_1/2}-1\right)\right]\frac{\left(\mathrm{e}^{-i\hat{x}_1}-1\right)\left(\mathrm{e}^{-i\hat{x}_2}-1\right)}{\hat{x}_1\hat{x}_2(-\hat{x}_1^2-\hat{x}_2^2)}\,\mathrm{d}\hat{x}_1\,\mathrm{d}\hat{x}_2\\ &= \frac{1}{24\,\pi}(2\pi+2\ln 2-11) = -0.04417. \end{aligned}$$

The total vector on the right-hand side is due to symmetry

$$F^T = (-0.04417,-0.04417,-0.04417,-0.04417,-0.04417,-0.04417,-0.04417,-0.04417).$$

The unknown coefficients for the boundary flux are obtained by $H^{-1}F$:

$$\mathrm{t}^T = (0.2564, 0.2564, 0.2564, 0.2564, 0.2564, 0.2564, 0.2564, 0.2564). \qquad (5.9)$$

As a first control, we take another Gauss theorem, the heat equilibrium condition. The integral of the fluxes at the boundary stands in equilibrium with the total sources ($L^i_e = 1/2$ is th length of the i-th element)

$$\sum_{i=1}^{8} L^i_e \mathrm{t}^i = 1.02544 \overset{!}{\approx} -\int_\Omega f(x)\,\mathrm{d}x = 1. \qquad (5.10)$$

The coarse mesh leads to an acceptable result, the global error is 2.5%.

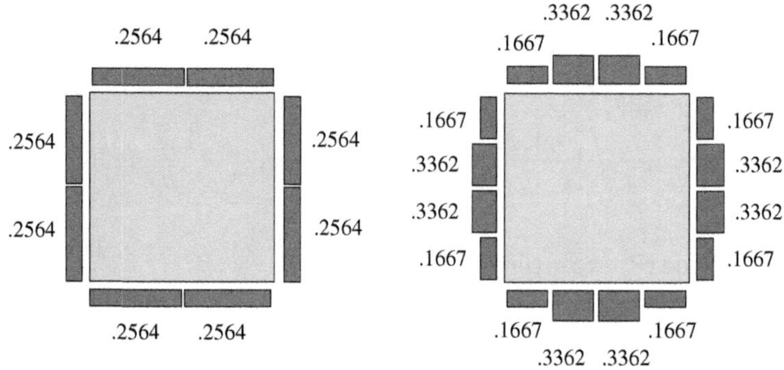

Figure 5.3: Flux at the boundary (left: 4 × 2 elements; right: 4 × 4 elements)

The results for a refined boundary mesh with four elements on each side are shown in the right part of Fig.5.3. The global error of the equilibrium between these boundary values and the sources f is about 0.6%.

$$\sum_{i=1}^{16} L_e^i t^i = 1.00597 \overset{!}{\approx} \int_\Omega f(x)\,dx = 1. \tag{5.11}$$

Finally, we compare the results with the exact solution expressed by Fourier series obtained by the separation of variables. With the sources $f(x) = 1$ for $x \in \Omega$ the temperature is obtained for all x from, cf. [GL65],

$$\begin{aligned} u(x) &= \sum_{k=1}^{\infty}\sum_{l=1}^{\infty} c_{kl} \sin(k\pi x_1)\sin(l\pi x_2); \qquad (5.12) \\ c_{kl} &= -\frac{4}{\pi^2(k^2+l^2)} \int_0^1\int_0^1 f(x_1,x_2)\sin(k\pi x_1)\sin(l\pi x_2)\,dx_1\,dx_2; \end{aligned}$$

and the flux at the boundary with the normal ν by

$$t(x) = -\nu \cdot \nabla u(x). \tag{5.13}$$

In the left graph of Fig.5.4 the results (the flux on one boundary) are given

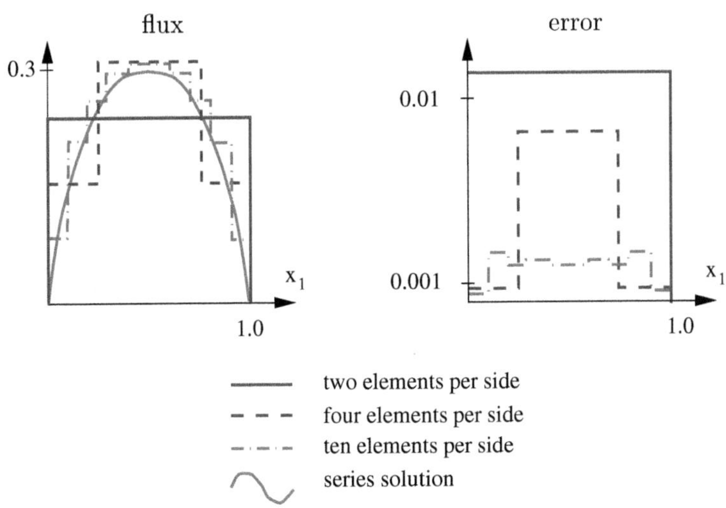

Figure 5.4: Result for the isotropic heat equation (2,4,10 elements per side)

for a discretization with 2, 4, or 10 elements per side and compared to the analytical solution (the series is truncated, i.e. the first 12 eigenfunctions in both directions are taken: $k = 1..12, l = 1..12$). In the right part, the mean error within each element (the difference between the two solutions) is plotted. A discretization of 10 elements per side is sufficent to reduce the local error to approximately 1%. A total number of $12^2 = 144$ eigenfunctions are related here to $4 \times 10 = 40$ boundary elements.

The postprocessing : interior values for the temperature

The general BIE for the evaluation of the interior values of the temperature are :

$$\begin{aligned}
\langle \phi_\Omega^j, u_\chi \rangle &= \langle \phi_\Omega^j, f_\chi * U \rangle + \sum_i^{N_t} t^i \langle \phi_\Omega^j, \phi_t^i * U \rangle \\
\overset{\mathcal{F}}{\leftrightarrow} \quad \left\langle \hat{\phi}_\Omega^j(-\hat{x}), \hat{u}_\chi \right\rangle &= \left\langle \hat{\phi}_\Omega^j(-\hat{x}), \hat{f}_\chi \hat{U} \right\rangle + \sum_i^{N_t} t^i \left\langle \hat{\phi}_\Omega^j(-\hat{x}), \hat{\phi}_t^i \hat{U} \right\rangle .
\end{aligned}$$

The simplest choice for the interior test functions ϕ_Ω^j, the Dirac-distributions $\delta(x - \xi^j)$, leads to a point-wise evaluation of the interior temperature $u(x = \xi^j)$ due to $\langle \delta(x - \xi^j), u_\chi(x) \rangle = u(\xi^j), \xi^j \in \Omega$:

$$\begin{aligned}
u(\xi^j) &= \langle \delta(x - \xi^j), f_\chi * U \rangle + \sum_i^{N_t} t^i \langle \delta(x - \xi^j), \phi_t^i * U \rangle \\
\overset{\mathcal{F}}{\leftrightarrow} \quad (2\pi)^2 u(\xi^j) &= \left\langle e^{-i<\hat{x},\xi^j>}, \hat{f}_\chi \hat{U} \right\rangle + \sum_i^{N_t} t^i \left\langle e^{-i<\hat{x},\xi^j>}, \hat{\phi}_t^i \hat{U} \right\rangle . \qquad (5.14)
\end{aligned}$$

Fig.5.5 (left) gives the result for the example of the coarsest mesh (4×2 elements). The error, i.e. the difference to the series solution obtained with with 12×12 eigenfunctions, for this very coarse mesh is already reasonable small (right graph of Fig.5.5). This means that a total of eight boundary elements already gives a result which differs by only 5% from an eigenfunction solution (144 eigenfunctions).

5.1.2 The Neumann problem

Next, the equivalence of the traditional approach and Fourier BEM is studied when singularities (in this case hypersingularities) are present. The rigorous distributional formulation of the BIE developed in Chapter 3 enables the accurate handling of all singularities which occur in the single integrals. It shows in total that all non-integrable singularities vanish if the terms are

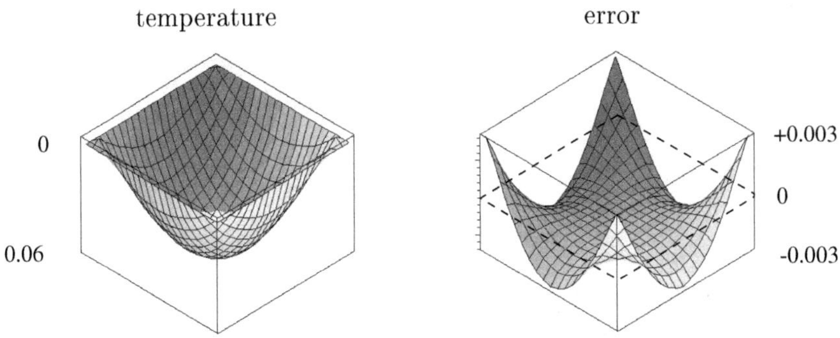

Figure 5.5: Result of the Dirichlet problem (2 elements per side); left : the temperature u in the interior; right : the difference to the series solution

combined adequately[1]. No regularization is needed either in the original space or in Fourier space.

Fourier-Galerkin boundary integral equations

As an example for a Neumann problem, we are considering

$$\begin{aligned} \Delta u(x) &= -f(x), & x \in \Omega; \\ t(x) &= t_\Gamma = 0, & x \in \Gamma, \end{aligned} \tag{5.15}$$

in the quadratic two-dimensional domain $\Omega = [0,1] \times [0,1]$. Here, the flux t at the boundaries is set to zero. The inner plane is heated as in Chapter 5.1.1. The symmetric Galerkin BIE for this Neumann problem is

$$\begin{aligned} \left\langle \phi_{\mathrm{u}}^j, \mathcal{A}_{\mathrm{t}}^j\{u_\chi\}\right\rangle &= \left\langle \phi_{\mathrm{u}}^j, f_\chi * \mathcal{A}_{\mathrm{t}}^j U\right\rangle - \sum_i^{N_{\mathrm{u}}} \mathrm{u}^i \left\langle \phi_{\mathrm{u}}^j, \phi_{\mathrm{u}}^i * \mathcal{A}_{\mathrm{t}}^j \mathcal{A}_{\mathrm{t}}^i U\right\rangle \qquad (5.16) \\ \overset{\mathcal{F}}{\leftrightarrow} \left\langle \hat{\phi}_{\mathrm{u}}^j(-\hat{x}), \hat{\mathcal{A}}_{\mathrm{t}}^j\{\hat{u}_\chi\}\right\rangle &= \left\langle \hat{\phi}_{\mathrm{u}}^j(-\hat{x}), \hat{f}_\chi \hat{\mathcal{A}}_{\mathrm{t}}^j \hat{U}\right\rangle - \sum_i^{N_{\mathrm{u}}} \mathrm{u}^i \left\langle \hat{\phi}_{\mathrm{u}}^j(-\hat{x}), \hat{\phi}_{\mathrm{u}}^i \hat{\mathcal{A}}_{\mathrm{t}}^j \hat{\mathcal{A}}_{\mathrm{t}}^i \hat{U}\right\rangle. \end{aligned}$$

The differential boundary operators are (ν^j, ν^i are the normal vectors of the test function ϕ_{u}^j and the trial functions ϕ_{u}^i, respectively)

$$\begin{aligned} \mathcal{A}_{\mathrm{t}}^j = \nu_k^j \, \partial_k \quad &\overset{\mathcal{F}}{\leftrightarrow} \quad \hat{\mathcal{A}}_{\mathrm{t}}^j = i\nu_k^j \, \hat{x}_k; \qquad (5.17) \\ \mathcal{A}_{\mathrm{t}}^i = \nu_k^i \, \partial_k \quad &\overset{\mathcal{F}}{\leftrightarrow} \quad \hat{\mathcal{A}}_{\mathrm{t}}^i = i\nu_k^i \, \hat{x}_k. \end{aligned}$$

[1]Compare Frangi & Guiggiani [GKRR92, FG01] and Mikhlin [MP86].

Hence, the fundamental flux is obtained as

$$\begin{aligned} T &= \mathcal{A}_t^j U = \nu^j \cdot \nabla U = \frac{\nu_1^j x_1 + \nu_2^j x_2}{2\pi(x_1^2 + x_2^2)} \qquad (5.18) \\ \overset{\mathcal{F}}{\leftrightarrow} \quad \hat{T} &= \hat{\mathcal{A}}_t^j \hat{U} = \nu^j \cdot i\hat{x}\hat{U} = \frac{i\nu_1^j \hat{x}_1 + i\nu_2^j \hat{x}_2}{-\hat{x}_1^2 - \hat{x}_2^2}, \end{aligned}$$

and as the hypersingular term (the explicit term in the original space is suppressed here, it is too lengthy)

$$\begin{aligned} S &= \mathcal{A}_t^j \mathcal{A}_t^i U = \nu^j \cdot \nabla(\nu^i \cdot \nabla U) \qquad (5.19) \\ \overset{\mathcal{F}}{\leftrightarrow} \quad \hat{S} &= \hat{\mathcal{A}}_t^j \hat{\mathcal{A}}_t^i \hat{U} = \frac{(\nu_1^j \hat{x}_1 + \nu_2^j \hat{x}_2)(\nu_1^i \hat{x}_1 + \nu_2^i \hat{x}_2)}{\hat{x}_1^2 + \hat{x}_2^2}. \end{aligned}$$

The left-hand side of (5.16) can be simplified ($u_\chi = \chi(x)u(x)$ and $\mathcal{A}_t^j u = t_\Gamma = 0$) :

$$\left\langle \phi_u^j, \mathcal{A}_t^j\{u_\chi\} \right\rangle = \left\langle \phi_u^j, (\chi \mathcal{A}_t^j u + u \mathcal{A}_t^j \chi) \right\rangle = \left\langle \phi_u^j, u \mathcal{A}_t^j \chi \right\rangle. \qquad (5.20)$$

This free term is singular; its singularity is of type $\delta(0)$ which will be shown in the next paragraph.

The singular integrals

The entries for the matrix are obtained from ($i, j = 1, 2, \ldots, N_u$)

$$G^{ji} = \left\langle \phi_u^j, \phi_u^i * \mathcal{A}_t^j \mathcal{A}_t^i U \right\rangle = \frac{1}{(2\pi)^2} \left\langle \hat{\phi}_u^j(-\hat{x}), \hat{\phi}_u^i \hat{\mathcal{A}}_t^j \hat{\mathcal{A}}_t^i \hat{U} \right\rangle \qquad (5.21)$$

Because of the hypersingularity of the kernel $\mathcal{A}_t^j \mathcal{A}_t^i U$ at $x = 0$ there are some pairs of test and trial functions which lead to infinite entries. The appropriate combination of these singularities, which is derived here, shows that finally all combined integrals are at most weakly singular and can be evaluated with standard routines. The only condition required is that continuous test and trial functions are chosen (they should belong at least to C^0). This continuity is already achieved by linear polynomials, cf. [Bon99].

An example for pairs for which the kernels are singular is visualized in Fig.5.6; the corresponding distributional notations are :

$$\begin{aligned} \phi_u^1 &= (1 - 2x_1)\mathrm{H}(x_1)\mathrm{H}(1 - 2x_1)\delta(x_2) \quad \overset{\mathcal{F}}{\leftrightarrow} \quad \hat{\phi}_u^1 = \hat{x}_1^{-2}\left(2 - i\hat{x}_1 - 2\mathrm{e}^{-i\hat{x}_1/2}\right); \\ \phi_u^2 &= (2x_1)\mathrm{H}(x_1)\mathrm{H}(1 - 2x_1)\delta(x_2) \quad \overset{\mathcal{F}}{\leftrightarrow} \quad \hat{\phi}_u^2 = \hat{x}_1^{-2}\left(i\mathrm{e}^{-i\hat{x}_1/2}(i\hat{x}_1 + 2) - 2\right); \\ \phi_u^3 &= (1 - 2x_2)\mathrm{H}(x_2)\mathrm{H}(1 - 2x_2)\delta(x_1) \quad \overset{\mathcal{F}}{\leftrightarrow} \quad \hat{\phi}_u^3 = \hat{x}_2^{-2}\left(2 - i\hat{x}_2 - 2\mathrm{e}^{-i\hat{x}_2/2}\right). \end{aligned}$$

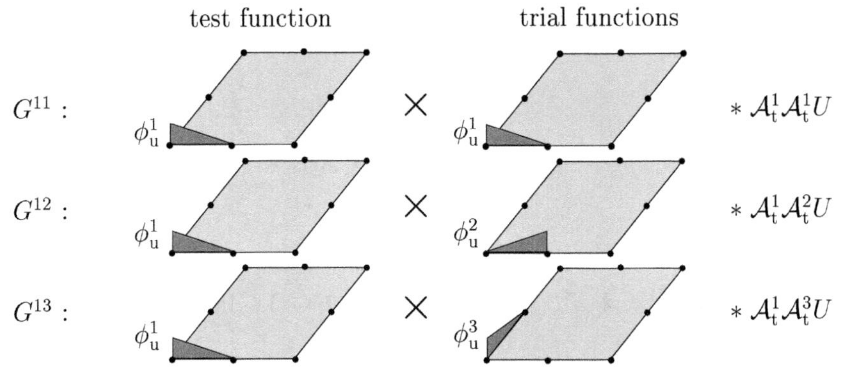

Figure 5.6: The pairs of trial and test functions with infinite matrix entries

In the original space we get for the first singular entry

$$
\begin{aligned}
G^{11} &= \left\langle \phi_{\mathrm{u}}^1, \phi_{\mathrm{u}}^1 * \mathcal{A}_{\mathrm{t}}^1\mathcal{A}_{\mathrm{t}}^1 U\right\rangle = \frac{1}{2\pi}\int_0^{1/2}(1-2x_1)\int_0^{1/2}\frac{1-2y_1}{(x_1-y_1)^2}\,\mathrm{d}y_1\,\mathrm{d}x_1 \\
&= \frac{1}{\pi}\int_0^{1/2}\left[((2x_1-1)(\ln|x_1-1/2|-\ln|x_1|))+1-\frac{1}{2x_1}\right]\mathrm{d}x_1 \\
&= \frac{1}{4\pi}(2\ln 2+1)+\lim_{\varepsilon\to 0}\frac{1}{2\pi}\ln\varepsilon+\frac{1}{6}\delta(0). \qquad (5.22)
\end{aligned}
$$

The first part of the last integral, i.e. $\int_0^{1/2}((2x_1-1)(\ln|x_1-1/2|-\ln|x_1|))\,\mathrm{d}x_1$ in (5.22), is weakly singular at $x_1 = 0$ and therefore in tegrable. The last term $\int_0^{1/2} 1/(2x_1)\,\mathrm{d}x_1$ is a homogeneous distribution, cf. [Hör90] which is only defined as a Hadamard finite part. The additional infinite part $\delta(0)/6$, which is normally neglected b y Hadamard's concept, was obtained b y an in tegration via distributional F ourier transform. This result is more easily obtained in the transformed space :

$$
\begin{aligned}
G^{11} &= \frac{1}{(2\pi)^2}\left\langle \hat{\phi}_{\mathrm{u}}^1(-\hat{x}), \hat{\phi}_{\mathrm{u}}^1\hat{\mathcal{A}}_{\mathrm{t}}^1\hat{\mathcal{A}}_{\mathrm{t}}^1\hat{U}\right\rangle \\
&= \frac{1}{(2\pi)^2}\int_{\mathbb{R}^2}\frac{\hat{x}_2^2\left(2+i\hat{x}_1-2\mathrm{e}^{i\hat{x}_1/2}\right)\left(2-i\hat{x}_1-2\mathrm{e}^{-i\hat{x}_1/2}\right)}{\hat{x}_1^4(\hat{x}_1^2+\hat{x}_2^2)}\,\mathrm{d}\hat{x}_1\,\mathrm{d}\hat{x}_2 \\
&= \frac{1}{4\pi}(2\ln 2+1)+\lim_{\varepsilon\to 0}\frac{1}{2\pi}\ln\varepsilon+\frac{1}{6}\delta(0). \qquad (5.23)
\end{aligned}
$$

The second singular value is

$$\begin{aligned} G^{12} &= \langle \phi_{\rm u}^1, \phi_{\rm u}^2 * \mathcal{A}_{\rm t}^1 \mathcal{A}_{\rm t}^2 U \rangle = -\frac{1}{4\pi} + \frac{1}{12}\delta(0) \\ \overset{\mathcal{F}}{\leftrightarrow} \quad G^{12} &= \frac{1}{(2\pi)^2} \left\langle \hat{\phi}_{\rm u}^1(-\hat{x}), \hat{\phi}_{\rm u}^2 \hat{\mathcal{A}}_{\rm t}^1 \hat{\mathcal{A}}_{\rm t}^2 \hat{U} \right\rangle = -\frac{1}{4\pi} + \frac{1}{12}\delta(0), \end{aligned} \tag{5.24}$$

and the third is

$$\begin{aligned} G^{13} &= \langle \phi_{\rm u}^1, \phi_{\rm u}^3 * \mathcal{A}_{\rm t}^j \mathcal{A}_{\rm t}^i U \rangle = -\frac{1}{8\pi}(6\ln 2 + 2 + \pi) - \lim_{\varepsilon\to 0} \frac{1}{2\pi} \ln \varepsilon \\ G^{13} &= \frac{1}{(2\pi)^2} \left\langle \hat{\phi}_{\rm u}^1(-\hat{x}), \hat{\phi}_{\rm u}^3 \hat{\mathcal{A}}_{\rm t}^j \hat{\mathcal{A}}_{\rm t}^i \hat{U} \right\rangle = -\frac{1}{8\pi}(6\ln 2 + 2 + \pi) - \lim_{\varepsilon\to 0} \frac{1}{2\pi} \ln \varepsilon. \end{aligned} \tag{5.25}$$

Another pair of two singularities originate from the free term (5.20). We have

$$\begin{aligned} \langle \phi_{\rm u}^1, u\mathcal{A}_{\rm t}^1 \chi \rangle &= -\langle (1-2x_1)\mathrm{H}(x_1)\mathrm{H}(1-2x_1)\delta(x_2), u(x_1,x_2)\delta(x_2) \rangle \\ &= -\delta(0) \int_0^{1/2} (1-2x_1) u(x_1, 0)\, \mathrm{d}x_1, \end{aligned}$$

where the local cutoff distribution $\chi = \mathrm{H}(x_2)$ and the local boundary differential operator $\mathcal{A}_{\rm t}^1 = \nu^1 \nabla = -\partial_2$ were taken, which results in $\mathcal{A}_{\rm t}^1 \chi = -\delta(x_2)$. Locally, $u(x_1, 0)$ was approximated by the polynomial ansatz functions on the boundary, i.e. by:

$$u(x_1, 0) \approx \mathrm{u}^1(1-2x_1) + \mathrm{u}^2(2x_1).$$

The free term on the right-hand side leads to a singular value related to u^1

$$\begin{aligned} R_{\rm u}^{11} &= -\langle (1-2x_1)\mathrm{H}(x_1)\mathrm{H}(1-2x_1)\delta(x_2), (1-2x_1)\delta(x_2) \rangle \\ &= \delta(0) \int_0^{1/2} (1-2x_1)^2\, \mathrm{d}x_1 = \frac{1}{6}\delta(0) \\ \overset{\mathcal{F}}{\leftrightarrow} \quad R_{\rm u}^{11} &= \frac{1}{(2\pi)^2} \left\langle \hat{\eta}_1(-\hat{x}), \hat{\mathcal{A}}_{\rm t}^1 \hat{\chi} \right\rangle = \frac{1}{6}\delta(0). \end{aligned} \tag{5.26}$$

and to another singular value related to u^2

$$\begin{aligned} R_{\rm u}^{12} &= -\langle (1-2x_1)\mathrm{H}(x_1)\mathrm{H}(1-2x_1)\delta(x_2), (2x_1)\delta(x_2) \rangle \\ &= \delta(0) \int_0^{1/2} (1-2x_1) 2x_1\, \mathrm{d}x_1 = \frac{1}{12}\delta(0) \\ \overset{\mathcal{F}}{\leftrightarrow} \quad R_{\rm u}^{12} &= \frac{1}{(2\pi)^2} \left\langle \hat{\eta}_2(-\hat{x}), \hat{\mathcal{A}}_{\rm t}^1 \hat{\chi} \right\rangle = \frac{1}{12}\delta(0). \end{aligned} \tag{5.27}$$

The easiest way to compute the free term in Fourier space is to rearrange the factors in the scalar product. We define :

$$\begin{aligned}
\bar{\phi}^1 &= (1-2x_1)^2 \mathrm{H}(x_1)\mathrm{H}(1-2x_1)\delta(x_2) \\
\overset{\mathcal{F}}{\leftrightarrow} \quad \hat{\bar{\phi}}^1 &= -\hat{x}_1^{-3}(i\hat{x}_1^2 - 4\hat{x}_1 - 8i + 8\mathrm{e}^{-i\hat{x}_1/2}); \\
\bar{\phi}^2 &= (1-2x_1)2x_1\mathrm{H}(x_1)\mathrm{H}(1-2x_1)\delta(x_2) \\
\overset{\mathcal{F}}{\leftrightarrow} \quad \hat{\bar{\phi}}^2 &= -\hat{x}_1^{-3}(2\hat{x}_1 + 8i + (2\hat{x}_1 - 8i)\mathrm{e}^{-i\hat{x}_1/2}).
\end{aligned}$$

Additionally we need $\hat{\mathcal{A}}_{\mathrm{t}}^1\chi = \mathcal{F}\{-\delta(x_2)\} = -2\pi\delta(\hat{x}_1)$.

These computations are presented explicitly to emphasize that all non-integrable singularities cancel. We have to combine

$$\begin{aligned}
\mathrm{u}^1(G^{11} + G^{13} - R_{\mathrm{u}}^{11}) &= -\frac{1}{8\pi}(2\ln 2 + \pi)\mathrm{u}^1; \\
\mathrm{u}^2(G^{12} - R_{\mathrm{u}}^{12}) &= -\frac{1}{4\pi}\mathrm{u}^2.
\end{aligned} \tag{5.28}$$

The final matrix $[G^{ji}]$ is of rank $N_{\mathrm{u}} - 1$, the problem is not unique.

5.1.3 The mixed boundary value problem

Compared to the Neumann problems, the more general case of mixed boundary value problems do not lead to additional singularities. The off-diagonal matrix has no singular entries. Therefore, we restrict the presentation to BIE and the explicit evaluation of the free terms.

The system of Galerkin boundary integral equations

The system of Galerkin BIE for the mixed boundary value problem

$$\begin{aligned}
\Delta u(x) &= -f(x), & x &\in \Omega; \\
u(x) &= u_\Gamma, & x &\in \Gamma_{\mathrm{u}}, \\
t(x) &= t_\Gamma, & x &\in \Gamma_{\mathrm{t}}
\end{aligned} \tag{5.29}$$

is in operator notation in the original space

$$\begin{aligned}
\langle \phi_{\mathrm{t}}^j, u_\chi \rangle &= \langle \phi_{\mathrm{t}}^j, f_\chi * U \rangle \qquad (5.30) \\
&- \sum_i^{N_{\mathrm{t}}} \mathrm{t}^i \langle \phi_{\mathrm{t}}^j, \phi_{\mathrm{t}}^i * U \rangle + \sum_i^{N_{\mathrm{u}}} \mathrm{u}^i \langle \phi_{\mathrm{t}}^j, \phi_{\mathrm{u}}^i * \mathcal{A}_{\mathrm{t}}^i U \rangle; \\
\langle \phi_{\mathrm{u}}^j, \mathcal{A}_{\mathrm{t}}^j\{u_\chi\} \rangle &= \langle \phi_{\mathrm{u}}^j, f_\chi * \mathcal{A}_{\mathrm{t}}^j U \rangle \qquad (5.31) \\
&+ \sum_i^{N_{\mathrm{t}}} \mathrm{t}^i \langle \phi_{\mathrm{u}}^j, \phi_{\mathrm{t}}^i * \mathcal{A}_{\mathrm{t}}^j U \rangle - \sum_i^{N_{\mathrm{u}}} \mathrm{u}^i \langle \phi_{\mathrm{u}}^j, \phi_{\mathrm{u}}^i * \mathcal{A}_{\mathrm{t}}^j \mathcal{A}_{\mathrm{t}}^i U \rangle,
\end{aligned}$$

and in Fourier space

$$\left\langle \hat{\phi}_{\mathrm{t}}^{j}(-\hat{x}), \hat{u}_{\chi} \right\rangle = \left\langle \hat{\phi}_{\mathrm{t}}^{j}(-\hat{x}), \hat{f}_{\chi}\hat{U} \right\rangle - \sum_{i}^{N_{\mathrm{t}}} \mathrm{t}^{i} \left\langle \hat{\phi}_{\mathrm{t}}^{j}(-\hat{x}), \hat{\phi}_{\mathrm{t}}^{i}\hat{U} \right\rangle + \sum_{i}^{N_{\mathrm{u}}} \mathrm{u}^{i} \left\langle \hat{\phi}_{\mathrm{t}}^{j}(-\hat{x}), \hat{\phi}_{\mathrm{u}}^{i}\hat{\mathcal{A}}_{\mathrm{t}}^{i}\hat{U} \right\rangle; \tag{5.32}$$

$$\left\langle \hat{\phi}_{\mathrm{u}}^{j}(-\hat{x}), \hat{\mathcal{A}}_{\mathrm{t}}^{j}\hat{u}_{\chi} \right\rangle = \left\langle \hat{\phi}_{\mathrm{u}}^{j}(-\hat{x}), \hat{f}_{\chi}\hat{\mathcal{A}}_{\mathrm{t}}^{j}\hat{U} \right\rangle + \sum_{i}^{N_{\mathrm{t}}} \mathrm{t}^{i} \left\langle \hat{\phi}_{\mathrm{u}}^{j}(-\hat{x}), \hat{\phi}_{\mathrm{t}}^{i}\hat{\mathcal{A}}_{\mathrm{t}}^{j}\hat{U} \right\rangle - \sum_{i}^{N_{\mathrm{u}}} \mathrm{u}^{i} \left\langle \hat{\phi}_{\mathrm{u}}^{j}(-\hat{x}), \hat{\phi}_{\mathrm{u}}^{i}\hat{\mathcal{A}}_{\mathrm{t}}^{j}\hat{\mathcal{A}}_{\mathrm{t}}^{i}\hat{U} \right\rangle, \tag{5.33}$$

The additional free terms

The left-hand side of the Galerkin displacement BIE (5.30) for a constant test function $\phi_{\mathrm{t}}^{1} = \mathrm{H}(x_1)\mathrm{H}(1-2x_1)\delta(x_2) \overset{\mathcal{F}}{\leftrightarrow} \hat{\phi}_{\mathrm{t}}^{1} = i\hat{x}_1^{-1}(\mathrm{e}^{-i\hat{x}_1/2} - 1)$ leads to

$$\langle \phi_{\mathrm{t}}^{1}, u_{\chi} \rangle = \int_{\mathbb{R}^2} \mathrm{H}(x_1)\mathrm{H}(1-2x_1)\delta(x_2)u(x)\chi(x)\,\mathrm{d}x = \frac{1}{2}\int_0^{1/2} u(x_1,0)\,\mathrm{d}x_1.$$

The relevant local part of the cutoff distribution χ for the domain Ω is equal to $\mathrm{H}(x_2)$. The kernel was simplified because of $\delta(x_2)\mathrm{H}(x_2) = \delta(x_2)/2$, cf. Chapter 3. With the local discretization $u(x_1,0) = \mathrm{u}^1(1-2x_1) + \mathrm{u}^2(2x_1)$ the free term is separated into two parts

$$R_{\mathrm{t}}^{11} = \frac{1}{2}\int_0^{1/2} (1-2x_1)\,\mathrm{d}x_1 = \frac{1}{8}; \qquad R_{\mathrm{t}}^{12} = \frac{1}{2}\int_0^{1/2} (2x_1)\,\mathrm{d}x_1 = \frac{1}{8}. \tag{5.34}$$

The free term of the traction BIE is according to (5.20)

$$\langle \phi_{\mathrm{u}}^{j}, \mathcal{A}_{\mathrm{t}}^{j}\{u_{\chi}\} \rangle = \langle \phi_{\mathrm{u}}^{j}, \chi\mathcal{A}_{\mathrm{t}}^{j}u \rangle + \langle \phi_{\mathrm{u}}^{j}, u\mathcal{A}_{\mathrm{t}}^{j}\chi \rangle.$$

The second term was discussed in (5.26) and (5.27). The first is equal to

$$\langle \phi_{\mathrm{u}}^{j}, \chi\mathcal{A}_{\mathrm{t}}^{j}u \rangle = \langle \phi_{\mathrm{u}}^{j}, \chi t \rangle. \tag{5.35}$$

For the first linear test function $(1-2x_1)\mathrm{H}(x_1)\mathrm{H}(1-2x_1)\delta(x_2)$ and the local approximation for the traction $t \approx \mathrm{t}^1 = $ const. we get

$$R_{\mathrm{t}}^{21} = R_{\mathrm{t}}^{12} = \frac{1}{2}\int_0^{1/2} (1-2x_1)\,\mathrm{d}x_1 = \frac{1}{8}. \tag{5.36}$$

Hence, there are no additional singularities. The same results for the free terms can be obtained in Fourier space.

5.1.4 Some computational aspects

In this section some comments on the analytical and numerical evaluation of the integrals in Fourier space are given. The Fourier kernels are in general highly oscillant and demand particular integration routines. Additionally, the integrals have to be evaluated over $\mathbb{R}^n$. A truncation necessary for numerical routines should be done carefully taking into account the fact that the kernels decrease in the worst cases by only $|\hat{x}|^{-2}$. The imaginary part of all kernels cancel.

Numerical computations in Fourier space

In the following, the numerical evaluation of one single matrix entry is discussed. As an example, we compute the value $\mathrm{u}^1(G^{11} + G^{13} - R_{\mathrm{u}}^{11})$ which was determined analytically in (5.28). The kernel leading to the singular free term R_{u}^{11} is concentrated on a line

$$\hat{\hat{\phi}}^1(-\hat{x})\hat{\mathcal{A}}_{\mathrm{t}}^1\hat{\chi} = -\frac{1}{3}\pi\delta(\hat{x}_1),$$

which is defined by a Dirac-distribution and hence cannot easily be represented numerically. It is the counterpart of the constant part of $G^{11} + G^{13}$ which is the singularity at infinity, i.e. we compute numerically (ν^1 is the normal vector corresponding to the test function)

$$\begin{aligned} I^{11} &= \frac{1}{(2\pi)^2}\int_{\mathbb{R}^n} (\hat{K}^{11} - \lim_{\nu^1\cdot\hat{x}\to\infty} \hat{K}^{11})\,\mathrm{d}\hat{x} = -0.18016, \\ \text{with}: \quad K^{11} &= \Re\left\{\hat{\phi}_{\mathrm{u}}^1(-\hat{x})\phi_{\mathrm{u}}^1(\hat{x})\hat{\mathcal{A}}_{\mathrm{t}}^1\hat{\mathcal{A}}_{\mathrm{t}}^1\hat{U}(\hat{x}) + \hat{\phi}_{\mathrm{u}}^1(-\hat{x})\phi_{\mathrm{u}}^3(\hat{x})\hat{\mathcal{A}}_{\mathrm{t}}^1\hat{\mathcal{A}}_{\mathrm{t}}^3\hat{U}(\hat{x})\right\}. \end{aligned}$$

The integral I^{11} is regular, it is neither locally nor globally singular and can be evaluated numerically, compare Fig.5.7,

$$\begin{aligned} \lim_{\hat{x}\to 0}(\hat{K}^{11} - \lim_{\nu^1\cdot\hat{x}\to\infty} \hat{K}^{11}) &= -\frac{1}{256\pi^2}; \\ \lim_{\hat{x}\to\infty}(\hat{K}^{11} - \lim_{\nu^1\cdot\hat{x}\to\infty} \hat{K}^{11}) &= \mathcal{O}\left(|\hat{x}|^{-2}\right). \end{aligned}$$

Analytic integrations in Fourier space

The analytic integration algorithms of the mathematical computer system MAPLE [Wat97] are not sufficiently capable of handling complicated expressions containing Dirac-, Heaviside- or signum-distributions. Whereas

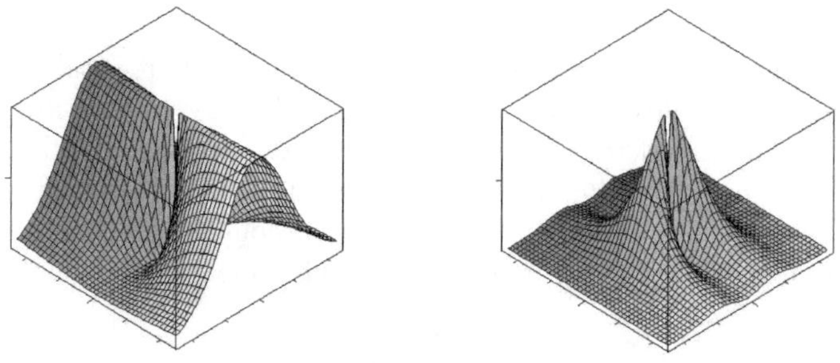

Figure 5.7: Left: The not combined singular kernel in Fourier space; right: The combined regular kernel for I^{11} in Fourier space

the "Fourier"-routine of the same program does not have these problems because it is based on the distributional definition of the Fourier transform, cf. [Dud97b]. Consequently the latter routine is used here to replace the required integrations. A standard integration over $\mathbb{R}^1$ is defined as

$$\int_{\mathbb{R}^n} \phi(y)\,\mathrm{d}y = \lim_{\hat{y}\to 0} \mathcal{F}\{\phi(y)\} = \lim_{\hat{y}\to 0} \int_{-\infty}^{+\infty} \phi(y)\,\mathrm{e}^{iy\hat{y}}\,\mathrm{d}y; \tag{5.37}$$

a successive application renders the multi-dimensional integrations.

5.2 Anisotropic Case

5.2.1 The mixed boundary value problem

The differential equation and its fundamental solution

The anisotropic heat conduction is described by the partial differential equation $-\mathrm{div}(K\nabla u) = f$ which is in tensor notation $(k, l = 1, 2)$

$$-\partial_k \left(K_{kl}\partial_l u\right) = f \qquad \overset{\mathcal{F}}{\leftrightarrow} \qquad -i\hat{x}_k \left(K_{kl} i\hat{x}_l \hat{u}\right) = \hat{f}. \tag{5.38}$$

The conductivity tensor $K = [K_{kl}]$ is a symmetric tensor of order two. Its eigenvectors point into the two principal directions of the conductivity. If K depends not on x, the differential operator is simply

$$\begin{aligned} \mathcal{P}(\partial) &= -K_{11}\partial_{11} - 2K_{12}\partial_{12} - K_{22}\partial_{22} \\ \overset{\mathcal{F}}{\leftrightarrow} \quad \hat{\mathcal{P}}(\hat{x}) &= K_{11}\hat{x}_1^2 + K_{12}\hat{x}_1\hat{x}_2 + K_{22}\hat{x}_2^2. \end{aligned} \tag{5.39}$$

The fundamental solution is hence ($|K| = \det K, \bar{r}^2 = x^T K^{-1} x$)

$$\begin{aligned} U &= -\frac{1}{4\pi}|K|^{-1/2} \ln \bar{r} \\ \overset{\mathcal{F}}{\leftrightarrow} \quad \hat{U} &= \frac{1}{K_{11}\hat{x}_1^2 + K_{12}\hat{x}_1\hat{x}_2 + K_{22}\hat{x}_2^2}. \end{aligned} \tag{5.40}$$

The term in the original space can be found for example in [Bon99]. The unknown flux at the boundary is obtained from

$$t = -\nu_k K_{kl} \partial_l u \qquad \overset{\mathcal{F}}{\leftrightarrow} \qquad \hat{t} = -i\nu_k K_{kl} \hat{x}_l \hat{u}. \tag{5.41}$$

Therefore, the other fundamental terms necessary for the Galerkin BIE are

$$\begin{aligned} \hat{T} &= \hat{\mathcal{A}}_{\mathrm{t}}^i \hat{U} = -i\nu_k^i K_{kl} \hat{x}_l \hat{U}; \\ \hat{S} &= \hat{\mathcal{A}}_{\mathrm{t}}^j \hat{\mathcal{A}}_{\mathrm{t}}^i \hat{U} = -\nu_k^i K_{kl} \hat{x}_l \nu_m^j K_{mn} \hat{x}_n \hat{U}, \end{aligned} \tag{5.42}$$

where ν^i, ν^j are the normal vectors of the i-th trial function and the j-th test function, respectively.

The system of Galerkin BIE in Fourier space

In analogy to the isotropic case, the Galerkin BIE are in Fourier space

$$\begin{aligned} \left\langle \hat{\phi}_{\mathrm{t}}^j(-\hat{x}), \hat{u}_\chi \right\rangle &= \left\langle \hat{\phi}_{\mathrm{t}}^j(-\hat{x}), \hat{f}_\chi \hat{U} \right\rangle + \sum_i^{N_{\mathrm{t}}} \mathrm{t}^i \left\langle \hat{\phi}_{\mathrm{t}}^j(-\hat{x}), \hat{\phi}_{\mathrm{t}}^i \hat{U} \right\rangle \\ &\quad - \sum_i^{N_{\mathrm{u}}} \mathrm{u}^i \left\langle \hat{\phi}_{\mathrm{t}}^j(-\hat{x}), \hat{\phi}_{\mathrm{u}}^i \hat{\mathcal{A}}_{\mathrm{t}}^i \hat{U} \right\rangle; \end{aligned} \tag{5.43}$$

$$\begin{aligned} -\left\langle \hat{\phi}_{\mathrm{u}}^j(-\hat{x}), \mathcal{A}_{\mathrm{t}}^j \hat{u}_\chi \right\rangle &= -\left\langle \hat{\phi}_{\mathrm{t}}^j(-\hat{x}), \hat{f}_\chi \hat{\mathcal{A}}_{\mathrm{t}}^j \hat{U} \right\rangle - \sum_i^{N_{\mathrm{t}}} \mathrm{t}^i \left\langle \hat{\phi}_{\mathrm{t}}^j(-\hat{x}), \hat{\phi}_{\mathrm{t}}^i \hat{\mathcal{A}}_{\mathrm{t}}^j \hat{U} \right\rangle \\ &\quad + \sum_i^{N_{\mathrm{u}}} \mathrm{u}^i \left\langle \hat{\phi}_{\mathrm{t}}^j(-\hat{x}), \hat{\phi}_{\mathrm{u}}^i \hat{\mathcal{A}}_{\mathrm{t}}^j \hat{\mathcal{A}}_{\mathrm{t}}^i \hat{U} \right\rangle. \end{aligned} \tag{5.44}$$

Attention should be paid to the free term $\mathcal{A}_{\mathrm{t}}^j \hat{u}_\chi$ as discussed in the isotropic case. All singularities cancel.

5.2.2 The Dirichlet problem

For the Dirichlet problem ($u = u_\Gamma = 0$) the system of BIE reduces to

$$0 = \left\langle \hat{\phi}_{\mathrm{t}}^j(-\hat{x}), \hat{f}_\chi \hat{U} \right\rangle + \sum_i^{N_{\mathrm{t}}} \mathrm{t}^i \left\langle \hat{\phi}_{\mathrm{t}}^j(-\hat{x}), \hat{\phi}_{\mathrm{t}}^i \hat{U} \right\rangle \tag{5.45}$$

The flux at the boundaries is approximated by 4×4 constant elements. The tensor of conductivity is chosen as

$$[K_{kl}] = \begin{bmatrix} 3 & 2 \\ 2 & 3 \end{bmatrix}; \qquad \xi_1 = \begin{pmatrix} 1 \\ -1 \end{pmatrix}, \xi_2 = \begin{pmatrix} 1 \\ 1 \end{pmatrix}. \tag{5.46}$$

The principal directions characterized by the eigenvectors $\xi_{1,2}$ of the medium lie in the diagonals of the cartesian coordinate system. The eigenvalues are $\omega_{1,2} = 1, 5$, i.e. the medium has in direction ξ_2 a conductivity which is five times higher than that in the other direction. The analytical integration of the first pair of trial and test functions, (ϕ^1, ϕ^1) from (5.5), leads to a result analogue to (5.6) :

$$H^{11} = \frac{1}{(2\pi)^2} \left\langle \hat{\phi}_t^1(-\hat{x}), \hat{\phi}_t^1 \hat{U} \right\rangle = -\frac{\sqrt{5}}{80\pi} (2 \ln 2 + 3) = -0.03903.$$

The explicit symmetric matrix is given here. It may serve as a benchmark,

$$H = \begin{bmatrix} -.03903 & -.01436 & -.00872 & -.00813 & -.00425 & .00006 & -.00145 & -.01559 \\ & -.03903 & -.02497 & -.00872 & .00006 & .00571 & .00546 & -.00145 \\ & & -.03903 & -.01436 & -.00145 & .00546 & .00571 & .00006 \\ & & & -.03903 & -.01559 & -.00145 & .00006 & -.00425 \\ & & & & -.03903 & -.01436 & -.00872 & -.00813 \\ & & \text{sym.} & & & -.03903 & -.02497 & -.00872 \\ & & & & & & -.03903 & -.01436 \\ & & & & & & & -.03903 \end{bmatrix}$$

In Fig.5.8 the temperature due to a single source in the center of the square is given. The anisotropic character is clearly reflected in this result.

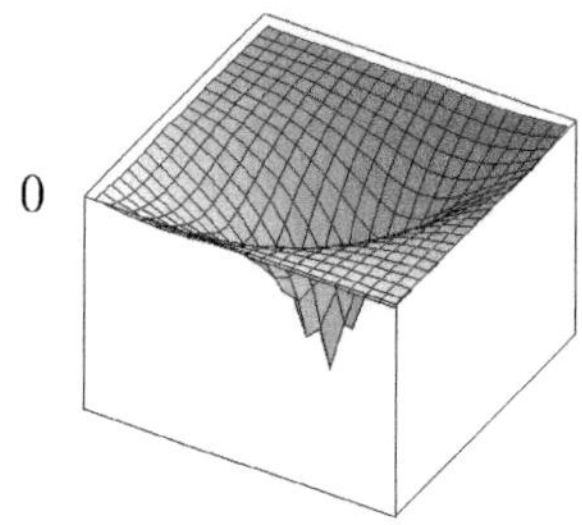

Figure 5.8: The temperature u due to a single centered source in an anisotropic medium (4×4 elements)

5.2.3 The three-dimensional case

The fundamental solution for the $\mathbb{R}^3$ is

$$U = \frac{1}{4\pi}|K|^{-1/2}\,\frac{1}{\bar{r}} \quad \overset{\mathcal{F}}{\leftrightarrow} \quad \hat{U} = -\,\frac{1}{K_{kl}\hat{x}_k\hat{x}_l}. \tag{5.47}$$

with $|K| = \det K, \bar{r}^2 = x^T K^{-1} x$ and $k, l = 1, 2, 3$.

Chapter 6

Elasticity

Claude L.M.H. Navier.

6.1 Isotropic Case

6.1.1 The mixed boundary value problem

Navier's differential equation and its fundamental solution

The strong form of Navier's differential equation for linear elastostatics is in three dimensions ($x, \hat{x} \in \mathbb{R}^3$)

$$\begin{aligned} \mu\, \partial_{ll} u_k + (\lambda + \mu)\, \partial_{kl} u_l &= -f_k, \qquad k,l = 1,2,3; \\ \overset{\mathcal{F}}{\leftrightarrow} \quad -\mu\, \hat{x}_l \hat{x}_l \hat{u}_k - (\lambda + \mu)\, \hat{x}_k \hat{x}_l \hat{u}_l &= -\hat{f}_k. \end{aligned} \tag{6.1}$$

u_k, f_k are the three components of the displacements and volume forces respectively. μ, λ are Lamé's constant parameters for the material which are related to Young's modulus E and the Poisson ratio $\bar{\nu}$ by

$$\lambda = \frac{E\bar{\nu}}{(1+\bar{\nu})(1-2\bar{\nu})}, \qquad \mu = \frac{E}{2(1+\bar{\nu})}. \tag{6.2}$$

We assume geometrical linearity (i.e. small displacements and small deformations) and physical linearity (Hooke's linear stress-strain relation). The fundamental solution is the response to a unit force in the direction x_m

$$f_k = \delta(x)\delta_{km} \qquad \overset{\mathcal{F}}{\leftrightarrow} \qquad \hat{f}_k = \delta_{km}. \tag{6.3}$$

δ_{km} is Kronecker's symbol. By rearranging (6.1) and inserting (6.3) the more suitable matrix form is obtained ($\mathcal{I}$ is the identity matrix)

$$\mathcal{P}(\partial)\, U = -\frac{1}{\rho}\delta(x)\mathcal{I} \qquad \overset{\mathcal{F}}{\leftrightarrow} \qquad \hat{\mathcal{P}}(\hat{x})\, \hat{U} = -\frac{1}{\rho}\mathcal{I}, \tag{6.4}$$

with the three-dimensional differential operator

$$\begin{aligned}
\mathcal{P}(\partial) &= \begin{bmatrix} c_s^2\Delta + c_1^2\partial_{11} & c_1^2\partial_{12} & c_1^2\partial_{13} \\ c_1^2\partial_{21} & c_s^2\Delta + c_1^2\partial_{22} & c_1^2\partial_{23} \\ c_1^2\partial_{31} & c_1^2\partial_{32} & c_s^2\Delta + c_1^2\partial_{33} \end{bmatrix} \\
\overset{\mathcal{F}}{\leftrightarrow} \quad \hat{\mathcal{P}}(\hat{x}) &= - \begin{bmatrix} c_s^2\hat{x}^2 + c_1^2\hat{x}_1^2 & c_1^2\hat{x}_1\hat{x}_2 & c_1^2\hat{x}_1\hat{x}_3 \\ c_1^2\hat{x}_2\hat{x}_1 & c_s^2\hat{x}^2 + c_1^2\hat{x}_2^2 & c_1^2\hat{x}_2\hat{x}_3 \\ c_1^2\hat{x}_3\hat{x}_1 & c_1^2\hat{x}_3\hat{x}_2 & c_s^2\hat{x}^2 + c_1^2\hat{x}_3^2 \end{bmatrix}
\end{aligned} \tag{6.5}$$

where the two parameters for the material are now the wave velocities c_p, c_s of the elastic medium ($c_1^2 = c_p^2 - c_s^2$)

$$c_p^2 = \frac{\lambda + 2\mu}{\rho}, \qquad c_s^2 = \frac{\mu}{\rho}. \tag{6.6}$$

The fundamental solution $\hat{U}$ in Fourier space is now derived by simple inversion from

$$\begin{bmatrix} c_s^2\hat{x}^2 + c_1^2\hat{x}_1^2 & c_1^2\hat{x}_1\hat{x}_2 & c_1^2\hat{x}_1\hat{x}_3 \\ c_1^2\hat{x}_2\hat{x}_1 & c_s^2\hat{x}^2 + c_1^2\hat{x}_2^2 & c_1^2\hat{x}_2\hat{x}_3 \\ c_1^2\hat{x}_3\hat{x}_1 & c_1^2\hat{x}_3\hat{x}_2 & c_s^2\hat{x}^2 + c_1^2\hat{x}_3^2 \end{bmatrix} \begin{bmatrix} \hat{U}_{11} & \hat{U}_{12} & \hat{U}_{13} \\ \hat{U}_{21} & \hat{U}_{22} & \hat{U}_{23} \\ \hat{U}_{31} & \hat{U}_{32} & \hat{U}_{33} \end{bmatrix} = \begin{bmatrix} 1 & 0 & 0 \\ 0 & 1 & 0 \\ 0 & 0 & 1 \end{bmatrix}$$

Hence $\hat{U}$ is

$$\hat{U}^{3\mathrm{d}} = [\hat{U}_{km}]^{3\mathrm{d}} = \begin{bmatrix} c_s^2\hat{x}^2 + c_1^2\hat{x}_1^2 & c_1^2\hat{x}_1\hat{x}_2 & c_1^2\hat{x}_1\hat{x}_3 \\ c_1^2\hat{x}_2\hat{x}_1 & c_s^2\hat{x}^2 + c_1^2\hat{x}_2^2 & c_1^2\hat{x}_2\hat{x}_3 \\ c_1^2\hat{x}_3\hat{x}_1 & c_1^2\hat{x}_3\hat{x}_2 & c_s^2\hat{x}^2 + c_1^2\hat{x}_3^2 \end{bmatrix}^{-1} \tag{6.7}$$

For the two-dimensional case the Fourier fundamental solution is simply

$$\hat{U}^{2\mathrm{d}} = [\hat{U}_{km}]^{2\mathrm{d}} = \begin{bmatrix} c_s^2\hat{x}^2 + c_1^2\hat{x}_1^2 & c_1^2\hat{x}_1\hat{x}_2 \\ c_1^2\hat{x}_2\hat{x}_1 & c_s^2\hat{x}^2 + c_1^2\hat{x}_2^2 \end{bmatrix}^{-1} \tag{6.8}$$

The fundamental solutions for the original space in two and three dimensions can be found in the literature, e.g. [GF97]. They are with $|x| = (\sum x_k^2)^{1/2}$

$$\begin{aligned}
U_{km}^{3\mathrm{d}} &= \frac{1}{16\pi\mu(1-\bar{\nu})|x|}\left[(3-4\bar{\nu})\delta_{km} + \partial_k|x|\partial_m|x|\right]; &(6.9)\\
U_{km}^{2\mathrm{d}} &= \frac{1}{8\pi\mu(1-\bar{\nu})}\left[-(3-4\bar{\nu})\ln|x|\delta_{km} + \partial_k|x|\partial_m|x|\right]. &(6.10)
\end{aligned}$$

The system of Galerkin BIE

The quantity dual to the boundary displacements, the traction vector t, is derived from the strain-displacement relation (compatibility)

$$\varepsilon_{kl} = \frac{1}{2}(\partial_k u_l + \partial_l u_k) \tag{6.11}$$

and the stress-strain relation

$$\sigma_{kl} = \lambda\, \varepsilon_{mm}\delta_{kl} + 2\mu\, \varepsilon_{kl} = \lambda\, \partial_m u_m \delta_{kl} + \mu\, (\partial_k u_l + \partial_l u_k). \tag{6.12}$$

The traction vector $t^i = \nu^i \cdot \sigma$ of the i-th boundary element with the normal ν^i is derived directly from the displacements u by

$$\begin{aligned} t^i_k = \nu^i_l \sigma_{kl} &= \lambda\, \nu^i_l \partial_m u_m \delta_{kl} + \mu\, \nu^i_l (\partial_k u_l + \partial_l u_k) \\ &= (\lambda \nu^i_k \partial_m + \mu \nu^i_m \partial_k) u_m + \mu \nu^i_m \partial_m u_k \\ \overset{\mathcal{F}}{\leftrightarrow} \quad \hat{t}^i_k = \nu^i_l \hat{\sigma}_{kl} &= i(\lambda \nu^i_k \hat{x}_m + \mu \nu^i_m \hat{x}_k)\hat{u}_m + i\mu\nu^i_m \hat{x}_m \hat{u}_k. \end{aligned} \tag{6.13}$$

Hence the fundamental traction $\hat{T} = \hat{\mathcal{A}}^i_{\mathrm{t}} \hat{U}$ is in Fourier space

$$\hat{T}^i_{kn} = i(\lambda \nu^i_k \hat{x}_m + \mu \nu^i_m \hat{x}_k)\hat{U}_{mn} + i\mu\nu^i_m \hat{x}_m \hat{U}_{kn}. \tag{6.14}$$

The boundary operator $\hat{\mathcal{A}}^i_{\mathrm{t}}$ is obtained by solving $\hat{T}^i = \hat{\mathcal{A}}^i_{\mathrm{t}} \hat{U}$, it is

$$\hat{\mathcal{A}}^i_{\mathrm{t}} = i \begin{bmatrix} c_s^2 \nu^i \cdot |\hat{x}| + c_1^2 \nu^i_1 \hat{x}_1 & c_2^2 \nu^i_1 \hat{x}_2 + c_s^2 \nu^i_2 \hat{x}_1 & c_2^2 \nu^i_1 \hat{x}_3 + c_s^2 \nu^i_3 \hat{x}_1 \\ c_2^2 \nu^i_2 \hat{x}_1 + c_s^2 \nu^i_1 \hat{x}_2 & c_s^2 \nu^i \cdot |\hat{x}| + c_1^2 \nu^i_2 \hat{x}_2 & c_2^2 \nu^i_2 \hat{x}_3 + c_s^2 \nu^i_3 \hat{x}_2 \\ c_2^2 \nu^i_3 \hat{x}_1 + c_s^2 \nu^i_1 \hat{x}_3 & c_2^2 \nu^i_3 \hat{x}_2 + c_s^2 \nu^i_2 \hat{x}_3 & c_s^2 \nu^i \cdot |\hat{x}| + c_1^2 \nu^i_3 \hat{x}_3 \end{bmatrix} \tag{6.15}$$

with $c_2^2 = c_p^2 - 2c_s^2 = \lambda/\rho$.

We can now establish the system of the vectorial Galerkin BIE in Fourier space for the displacements and the tractions:

$$\begin{aligned} \left\langle \hat{\phi}^j_{\mathrm{t}}(-\hat{x}), \hat{u}_\chi \right\rangle &= \left\langle \hat{\phi}^j_{\mathrm{t}}(-\hat{x}), \hat{f}_\chi \hat{U} \right\rangle \\ &\quad + \sum_i^{N_{\mathrm{t}}} \mathfrak{t}^i \left\langle \hat{\phi}^j_{\mathrm{t}}(-\hat{x}), \hat{\phi}^i_{\mathrm{t}} \hat{U} \right\rangle - \sum_i^{N_{\mathrm{u}}} \mathfrak{u}^i \left\langle \hat{\phi}^j_{\mathrm{t}}(-\hat{x}), \hat{\phi}^i_{\mathrm{u}} \hat{\mathcal{A}}^i_{\mathrm{t}} \hat{U} \right\rangle; \end{aligned} \tag{6.16}$$

$$\begin{aligned} -\left\langle \hat{\phi}^j_{\mathrm{u}}(-\hat{x}), \hat{\mathcal{A}}^j_{\mathrm{t}} \hat{u}_\chi \right\rangle &= -\left\langle \hat{\phi}^j_{\mathrm{u}}(-\hat{x}), \hat{f}_\chi \hat{\mathcal{A}}^j_{\mathrm{t}} \hat{U} \right\rangle \\ &\quad - \sum_i^{N_{\mathrm{t}}} \mathfrak{t}^i \left\langle \hat{\phi}^j_{\mathrm{u}}(-\hat{x}), \hat{\phi}^i_{\mathrm{t}} \hat{\mathcal{A}}^j_{\mathrm{t}} \hat{U} \right\rangle + \sum_i^{N_{\mathrm{u}}} \mathfrak{u}^i \left\langle \hat{\phi}^j_{\mathrm{u}}(-\hat{x}), \hat{\phi}^i_{\mathrm{u}} \hat{\mathcal{A}}^j_{\mathrm{t}} \hat{\mathcal{A}}^i_{\mathrm{t}} \hat{U} \right\rangle. \end{aligned} \tag{6.17}$$

In an abbreviated notation this is :

$$\left\langle \hat{\phi}^j(-\hat{x}), \hat{\mathcal{B}} \hat{u}_\chi \right\rangle = \left\langle \hat{\phi}^j(-\hat{x}), \hat{f}_\chi \hat{\mathcal{B}} \hat{U} \right\rangle + \sum_i \mathfrak{v}^i \left\langle \hat{\phi}^j(-\hat{x}), \hat{\phi}^i \hat{\mathcal{A}} \hat{U} \right\rangle, \tag{6.18}$$

with the operators

$$\hat{\mathcal{A}} = \begin{bmatrix} \mathcal{I} & -\mathcal{A}_t^j \\ -\mathcal{A}_t^j & \mathcal{A}_t^j \mathcal{A}_t^i \end{bmatrix}, \qquad \hat{\mathcal{B}} = \begin{pmatrix} \mathcal{I} \\ -\mathcal{A}_t^j \end{pmatrix}. \tag{6.19}$$

$\mathcal{I}$ is the identity operator. This schematic approach facilitates the transfer to other physical problems. Note that all quantities in these equations are either vectors or matrices.

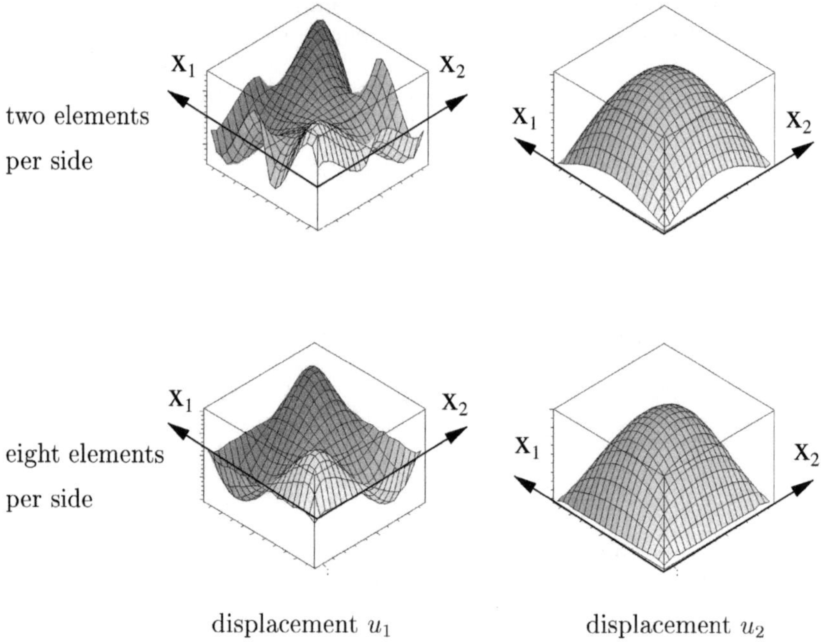

Figure 6.1: The displacements u_1, u_2 in the interior for the Dirichlet problem of isotropic elasticity

6.1.2 The Dirichlet problem in two dimensions

The two-dimensional Dirichlet problem, i.e. fixed boundaries,

$$\begin{aligned} \mu\,\partial_{ll}u_k + (\lambda+\mu)\,\partial_{kl}u_l &= f_k, && x \in \Omega \subset \mathbb{R}^2; \\ u_k &= 0, && x \in \Gamma. \end{aligned} \tag{6.20}$$

finds its equivalent in the Galerkin BIE obtained from (6.16)

$$0 = \left\langle \hat{\phi}_t^j, \hat{f}_x \hat{U} \right\rangle + \sum_i^{N_t} \mathrm{t}^i \left\langle \hat{\phi}_t^j, \hat{\phi}_t^i \hat{U} \right\rangle . \tag{6.21}$$

The indices i, j run over the total number of degrees of freedom N_t.

As example, a simple square $\Omega = [0,1] \times [0,1]$ is studied. The material is described by Poisson's ratio $\bar{\nu} = 1/3$ and Young's modulus $E = 2.1 \cdot 10^5$ [MPa]. The tractions $t = (t_1, t_2)$ on the boundaries are approximated by constant trial functions ϕ_{t1}^i, ϕ_{t2}^i. In the first case two elements and in the second eight elements were chosen for each side. Hence the total number of degrees of freedom are $N_t = 2 \times 4 \times 2 = 16$ and $N_t = 2 \times 4 \times 8 = 64$. As volume force a constant uniform loading in x_2 direction is taken, i.e. $f_1 = 0, f_2 = 1$ for $x \in \Omega$. The results for the displacements $u = (u_1, u_2)$ in the interior are given in Fig.6.1. The boundary conditions $u_1 = u_2 = 0$ are still only very approximatively fulfilled in the case of two elements per side, though sufficiently well by the finer model with eight elements on each side.

6.2 Anisotropic Case

6.2.1 The state of the art

For general anisotropy, it is not possible to find an analytic expression for the fundamental solution U in the original space. The approaches which can be found in the literature, e.g. [Kög00, SG98, Sch94b], are based on the numerical evaluation of u. Unfortunately this is too time consuming to be combined directly within a standard BEM algorithm. Therefore, a sufficiently complete table of numerical evaluated values of the fundamental solution has to be established in advance, cf. [WC78].The actual BEM problem is then solved by cubic interpolation from these tabulated precalculated values. The main drawbacks are the high storage necessary for the tabulated values and the dependence of the accuracy on the degree of anisotropy. The fundamental solution must be sufficiently smooth which is only the case in static problems. The interpolation remains doubtful in the vicinity of the singularity.

Alternatively, the values of the fundamental solution are constructed from the isotropic case modified by a perturbation series, cf. [GGK96]. This method is only adequate for weak anisotropy. Another approach splits the integrand into the isotropic and the anisotropic part. The latter is handled like a volume term, for example via the dual reciprocity method [PBW92]. The choice of the appropriate number and location of internal points remains difficult.

The method proposed by [Nis88, SG98] is based upon the calculus of the residues of the inverse symbol. Nishimura's approach is also based on the Fourier fundamental solution but not on Parseval's theorem. The main problem seems to be the numerical location of the roots and the following numerical integration.

6.2.2 General anisotropic elasticity

The Galerkin BIE for general anisotropy

The stress tensor for general anisotropic media is, cf. [Kög00],

$$\sigma_{kl} = C_{klmn}\varepsilon_{mn}; \qquad \varepsilon_{mn} = \frac{1}{2}\left(\partial_n u_m + \partial_m u_n\right). \tag{6.22}$$

The 81 components of the elasticity tensor C_{klmn} can be deduced in the general case from 21 different material parameters. The equilibrium equation

$$\partial_l \sigma_{kl} = -f_k \tag{6.23}$$

leads with $f_k = \delta_{kp}\delta(x)$ to the partial differential equation for the determination of the fundamental solution U

$$C_{klmn}\partial_{nl}U_{pm} = -\delta_{kp}\delta(x) \qquad \overset{\mathcal{F}}{\leftrightarrow} \qquad C_{klmn}\hat{x}_n\hat{x}_l\hat{U}_{pm} = \delta_{kp}. \tag{6.24}$$

As will be shown in the examples, the evaluation of the fundamental solution in Fourier space is straightforward. The fundamental tractions at a boundary with the normal ν^i are given by

$$T^i_{pk} = C_{klmn}\partial_n U_{pm}\nu^i_l \qquad \overset{\mathcal{F}}{\leftrightarrow} \qquad \hat{T}^i_{pk} = iC_{klmn}\,\hat{x}_n\hat{U}_{pm}\nu^i_l, \tag{6.25}$$

which leads to the definition of the boundary traction operator

$$T^i = \mathcal{A}^i_t U \qquad \overset{\mathcal{F}}{\leftrightarrow} \qquad \hat{T}^i = \hat{\mathcal{A}}^i_t\hat{U}. \tag{6.26}$$

Finally, the formal notation for the Galerkin BIE is in analogy to (6.18)

$$\left\langle \hat{\phi}^j(-\hat{x}), \hat{\mathcal{B}}\hat{u}_\chi \right\rangle = \left\langle \hat{\phi}^j(-\hat{x}), \hat{f}_\chi \hat{\mathcal{B}}\hat{U} \right\rangle + \sum_i \mathfrak{v}^i \left\langle \hat{\phi}^j(-\hat{x}), \hat{\phi}^i \hat{\mathcal{A}}\hat{U} \right\rangle, \tag{6.27}$$

with $\hat{\mathcal{A}}, \hat{\mathcal{B}}$ as defined in (6.19).

6.2.3 Orthotropic elasticity in $\mathbb{R}^2$

The fundamental displacement and the fundamental traction

For orthotropic materials in the plane strain case, the relevant components of the elasticity tensor C_{klmn} are $C_{1111}, C_{1122}, C_{2222}, C_{1212}$. The two equations in (6.22) lead to, cf. [GP97],

$$\begin{aligned}\sigma_{11} &= C_{1111}\partial_1 u_1 + C_{1122}\partial_2 u_2; \\ \sigma^{12} &= \frac{1}{2}C_{1212}(\partial_2 u_1 + \partial_1 u_2); \\ \sigma^{22} &= C_{1122}\partial_1 u_1 + C_{2222}\partial_2 u_2.\end{aligned} \tag{6.28}$$

Inserted in (6.23), the differential operator is obtained as

$$\begin{aligned}\mathcal{P}(\partial) &= \begin{bmatrix} C_{1111}\partial_{11} + \frac{1}{2}C_{1212}\partial_{22} & (C_{1122} + \frac{1}{2}C_{1212})\partial_{12} \\ (C_{1122} + \frac{1}{2}C_{1212})\partial_{12} & C_{2222}\partial_{22} + \frac{1}{2}C_{1212}\partial_{11} \end{bmatrix} \\ \overset{\mathcal{F}}{\leftrightarrow} \quad \hat{\mathcal{P}}(\hat{x}) &= -\begin{bmatrix} C_{1111}\hat{x}_1^2 + \frac{1}{2}C_{1212}\hat{x}_2^2 & (C_{1122} + \frac{1}{2}C_{1212})\hat{x}_1\hat{x}_2 \\ (C_{1122} + \frac{1}{2}C_{1212})\hat{x}_2\hat{x}_1 & C_{2222}\hat{x}_2^2 + \frac{1}{2}C_{1212}\hat{x}_1^2 \end{bmatrix}.\end{aligned} \tag{6.29}$$

The fundamental solution in Fourier space is $\hat{U} = \mathcal{P}^{-1}(\hat{x})$

$$\hat{U} = -\frac{1}{N}\begin{bmatrix} C_{2222}\hat{x}_2^2 + \frac{1}{2}C_{1212}\hat{x}_1^2 & -(C_{1122} + \frac{1}{2}C_{1212})\hat{x}_1\hat{x}_2 \\ -(C_{1122} + \frac{1}{2}C_{1212})\hat{x}_2\hat{x}_1 & C_{1111}\hat{x}_1^2 + \frac{1}{2}C_{1212}\hat{x}_2^2 \end{bmatrix}, \tag{6.30}$$

with the determinant

$$N = (C_{1111}\hat{x}_1^4 - 2C_{1122}\hat{x}_1^2\hat{x}_2^2 + C_{2222}\hat{x}_2^4)C_{1212} + \hat{x}_1^2\hat{x}_2^2(C_{1111}C_{2222} - C_{1122}^2).$$

For this simple anisotropy, it is possible to derive a fundamental solution even in the original space[1]. It is here taken from [GP97, CS83]

$$U = \frac{1}{\beta}\begin{bmatrix} \alpha_1^{1/2}A_2^2\ln r_1 - \alpha_2^{1/2}A_1^2\ln r_2 & A_1A_2(\theta_2 - \theta_1) \\ \text{sym.} & \alpha_2^{-1/2}A_2^2\ln r_2 - \alpha_1^{-1/2}A_1^2\ln r_1 \end{bmatrix} \tag{6.31}$$

For the computation of these expressions in the original space the flexibility matrix s is needed

$$[s_{kl}] = \begin{bmatrix} C_{1111} & C_{1122} & 0 \\ C_{1122} & C_{2222} & 0 \\ 0 & 0 & \frac{1}{2}C_{1212} \end{bmatrix}^{-1} \tag{6.32}$$

[1] A.E. Green: A note on stress systems in aelotropic materials. *Phil. Mag.* **34**, 416–418 (1943).

The constants $\alpha_k, k = 1, 2$ are determined by

$$\alpha_1 + \alpha_2 = \frac{2s_{12} + s_{33}}{s_{22}} \qquad \text{and} \qquad \alpha_1 \alpha_2 = \frac{s_{11}}{s_{22}}.$$

Additionally we need ($k = 1, 2$)

$$\beta = 2\pi(\alpha_1 - \alpha_2)s_{22}; \qquad A_k = s_{12} - \alpha_k s_{22}; \qquad r_k = \sqrt{x_1^2 + \frac{1}{\alpha_k} x_2^2};$$

$$\theta_k = \arctan\left[\frac{x_2}{\sqrt{\alpha_k} x_1}\right].$$

In Fourier space the fundamental traction $\hat{T}$ is obtained by applying the traction operator

$$\hat{\mathcal{A}}_{\rm t}^i = i \begin{bmatrix} C_{1111}\hat{x}_1 \nu_1^i + \frac{1}{2} C_{1212}\hat{x}_2 \nu_2^i & C_{1122}\hat{x}_2 \nu_1^i + \frac{1}{2} C_{1212}\hat{x}_1 \nu_2^i \\ C_{1122}\hat{x}_1 \nu_2^i + \frac{1}{2} C_{1212}\hat{x}_2 \nu_1^i & C_{2222}\hat{x}_2 \nu_2^i + \frac{1}{2} C_{1212}\hat{x}_1 \nu_1^i \end{bmatrix} \tag{6.33}$$

to the fundamental solution $\hat{U}$, by simply multiplying the two matrices $\hat{\mathcal{A}}_{\rm t}^i$ and $\hat{U}$. Note that $\hat{\mathcal{A}}_{\rm t}^i$ is not symmetric. The lengthy expressions for the original space are omitted here, they can be found for example in [GP97].

The Galerkin BIE for orthotropic elasticity in $\mathbb{R}^2$ is finally obtained by inserting the derived Fourier fundamental solution $\hat{U}$ and the corresponding traction operator $\hat{\mathcal{A}}_{\rm t}^i$ in (6.16) and (6.17).

The clamped spruce board

As cited in [CS83] the behaviour of a spruce board with its grain parallel to the x_3 axis may be modelled by the model proposed in this section. The inversion of the flexibility matrix

$$[s_{kl}] = \begin{bmatrix} 15.500 & -0.330 & 0 \\ -0.330 & 0.587 & 0 \\ 0 & 0 & 11.500 \end{bmatrix} \times 10^{-4} \quad [\text{kPa}^{-1}].$$

leads to the elasticity coefficients, cf. (6.32),

$$\begin{bmatrix} C_{1111} & C_{1122} & 0 \\ C_{1122} & C_{2222} & 0 \\ 0 & 0 & \frac{1}{2}C_{1212} \end{bmatrix} = \begin{bmatrix} 0.0653 & 0.0367 & 0 \\ 0.0367 & 1.7242 & 0 \\ 0 & 0 & 0.0870 \end{bmatrix} \times 10^4 [\text{kPa}].$$

Inserted in (6.30) the Fourier fundamental solution is obtained easily. As sample Dirichlet problem the square $[0, 1] \times [0, 1]$ is taken charged with a single Dirac-load in the direction of x_2 at the center $x_1 = x_2 = 1/2$. The resulting displacements in the interior are given in Fig.6.2 and reflect clearly the anisotropic stiffness. The singularity of u_2 at the load point $x = (1/2, 1/2)$ is correctly represented (the value is missing in the graphic representation).

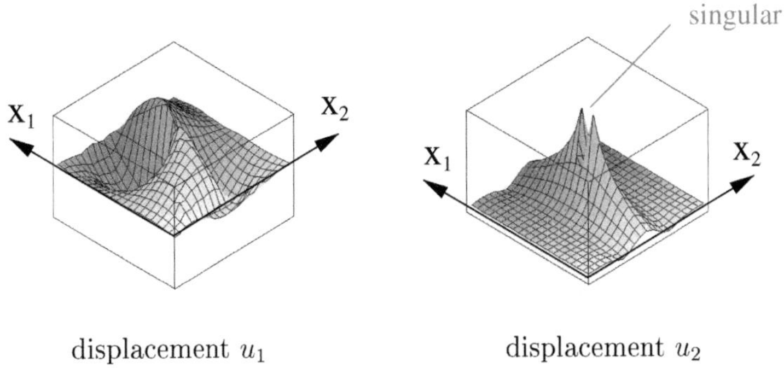

Figure 6.2: u_1, u_2 in the interior for the Dirichlet problem of transverse isotropic elasticity

6.2.4 Transverse isotropic elasticity in $\mathbb{R}^3$

The fundamental solution

In three dimensions, an analytic fundamental solution in the original space can only be found for transverse isotropic elasticity, cf. [Nis88]. If the plane of isotropy lies in the (x_1, x_2)-plane the relevant values of the elasticity tensor C_{klmn} are [GZ60]

$$\begin{array}{cccccc} C_{1111} & C_{1122} & C_{1133} & 0 & 0 & 0 \\ & C_{1111} & C_{1133} & 0 & 0 & 0 \\ & & C_{3333} & 0 & 0 & 0 \\ & & & C_{1313} & 0 & 0 \\ & & & & C_{1313} & 0 \\ & & & & & \frac{1}{2}(C_{1111} - C_{1122}) \end{array} \tag{6.34}$$

where the transverse isotropy implies

$$C_{2233} = C_{1133}, \quad C_{2323} = C_{1313}, \quad C_{2222} = C_{1111}, \quad C_{1212} = \frac{1}{2}(C_{1111} - C_{1122}).$$

We use the common abbreviations $C_{11} = C_{1111}, C_{33} = C_{3333}, C_{13} = C_{1133}, C_{44} = C_{1313}, C_{66} = C_{1212}$ and derive the transformed differential operator from (6.24)

$$\hat{\mathcal{P}}(\hat{x}) = \begin{bmatrix} C_{66}\hat{r}^2 + C_{44}\hat{x}_3^2 + C_1\hat{x}_1^2 & C_1\hat{x}_1\hat{x}_2 & C_2\hat{x}_1\hat{x}_3 \\ C_1\hat{x}_2\hat{x}_1 & C_{66}\hat{r}^2 + C_{44}\hat{x}_3^2 + C_1\hat{x}_2^2 & C_2\hat{x}_2\hat{x}_3 \\ C_2\hat{x}_3\hat{x}_1 & C_2\hat{x}_3\hat{x}_2 & C_{44}\hat{r}^2 + C_{33}\hat{x}_3^2 \end{bmatrix}$$

with $C_1 = C_{11} - C_{66}, C_2 = C_{13} + C_{44}$ and $\hat{r} = \sqrt{\hat{x}_1^2 + \hat{x}_2^2}$. Hence the Fourier fundamental solutions is

$$\hat{U} = \begin{bmatrix} C_{66}\hat{r}^2 + C_{44}\hat{x}_3^2 + C_1\hat{x}_1^2 & C_1\hat{x}_1\hat{x}_2 & C_2\hat{x}_1\hat{x}_3 \\ C_1\hat{x}_2\hat{x}_1 & C_{66}\hat{r}^2 + C_{44}\hat{x}_3^2 + C_1\hat{x}_2^2 & C_2\hat{x}_2\hat{x}_3 \\ C_2\hat{x}_3\hat{x}_1 & C_2\hat{x}_3\hat{x}_2 & C_{44}\hat{r}^2 + C_{33}\hat{x}_3^2 \end{bmatrix}^{-1} \tag{6.35}$$

Nishimura [Nis88] has derived an analytical expression in the original space, compare also Kröner [Krö53],

$$\begin{aligned} U_{\alpha\beta} &= \frac{1}{4\pi C_{66}}\left(\frac{|x_3|}{r_1\bar{r}_1}\delta_{\alpha\beta} + \frac{\kappa_1 x_\alpha x_\beta}{r_1\bar{r}_1^2}\right) + \sum_{k=2}^{3}\frac{K_k}{4\pi C_{44}}\left(\frac{\delta_{\alpha\beta}}{\bar{r}_k} - \frac{\kappa_k x_\alpha x_\beta}{r_k\bar{r}_k^2}\right), \quad (6.36) \\ U_{\alpha 3} &= U_{3\alpha} = \frac{\mathrm{sgn} x_3}{4\pi}\sum_{k=2}^{3}\frac{L_k x_\alpha}{C_{44} r_k \bar{r}_k}, \qquad U_{33} = \frac{1}{4\pi}\sum_{k=2}^{3}\frac{M_k}{C_{44} r_k}. \end{aligned}$$

We need $\kappa_1 = C_{44}/C_{66}$ and $\kappa_{2,3}$ as the two distinct roots of

$$C_{44}\kappa^2 - (C_{11}C_{33} - C_{13}^2 - 2C_{13}C_{44})\kappa + C_{33} = 0,$$

and $r_k = \sqrt{\kappa_k(x_1^2 + x_2^2) + x_3^3}, \bar{r}_k = r_k + |x_3|$. Finally the following constants have to be determined

$$K_2 = \frac{\kappa_2 C_{44} - C_{33}}{C_{11}(\kappa_2 - \kappa_3)}, \quad L_2 = \frac{(C_{13} + C_{44})\kappa_2}{C_{11}(\kappa_2 - \kappa_3)}, \quad M_2 = \frac{C_{11}\kappa_2 - C_{44}}{C_{11}(\kappa_2 - \kappa_3)}.$$

The values for K_3, L_3, M_3 can be obtained by interchanging κ_2 and κ_3.

Chapter 7

Plates

Gustav Robert Kirchhoff

7.1 The Thin Plate

In this chapter Fourier BEM is applied to plate problems. In the first part, the model for the bending of plates is analysed which is due to Kirchhoff[1] and which is valid for thin plates where shear deformations can be neglected. For this theory, collocation approaches are well established for the original space, e.g. [AP92, Bes91]. The Galerkin BIE were only recently developed in [FB98]. In the second part, we consider plates on Winkler foundations for which a collocation approach can be found in [Jah98]. Finally, the refined plate theory of Reissner is discussed, cf. [AP92, Bes91].

All these approaches were developed for isotropic plates in the original space. Hence, the Galerkin BIE for the anisotropic case is newly derived here in the original as well as in the transformed space.

7.1.1 Isotropic case

The differential equation and its fundamental solution

The midsurface of the plate is situated in the (x_1, x_2) plane. $w(x); x = x_1, x_2$ denotes the out-of-plane bending displacement. We define the unit outward normal $\nu = (\nu_1, \nu_2)^T$ and the unit tangent $\tau = (\tau_1, \tau_2)^T = (-\nu_2, \nu_1)^T$. The moment m_{kl} and the shear components q_k are constructed from the stress

[1]Kirchhoff. G.: Über das Gleichgewicht und die Bewegung einer elastischen Scheibe. *J. Reine und Angewandte Mathematik* **40** (1850), 51–88.

tensor σ_{kl} by

$$m_{kl} = \int_{-h/2}^{h/2} \sigma_{kl}(x, x_3) x_3 \, dx_3, \qquad q_k = \int_{-h/2}^{h/2} \sigma_{k3}(x, x_3) \, dx_3. \tag{7.1}$$

They can be related to w by

$$\begin{aligned} m_{kl} &= -K_{klmn}\partial_{mn}w & \overset{\mathcal{F}}{\leftrightarrow} \quad \hat{m}_{kl} &= K_{klmn}\hat{x}_m\hat{x}_n\hat{w}; \\ q_k &= \partial_l m_{kl} = -D\partial_{kll}w & \overset{\mathcal{F}}{\leftrightarrow} \quad \hat{q}_k &= i\hat{x}_l\hat{m}_{kl} = iD\hat{x}_k\hat{x}_l\hat{x}_l\hat{w}; \end{aligned} \tag{7.2}$$

$$\text{with } K_{klmn} = D[(1-\bar{\nu})\delta_{km}\delta_{ln} + \bar{\nu}\delta_{kl}\delta_{mn}], \qquad D = \frac{Eh^3}{12(1-\bar{\nu}^2)},$$

where summation has to be applied for repeated indices $k, l, m, n = 1, 2$. $E, \bar{\nu}, D$ are the Young's modulus of elasticity, the Poisson's coefficient, and the flexural rigidity of the plate, respectively. δ_{kl} is Kronecker's symbol, and h is the thickness of the plate.

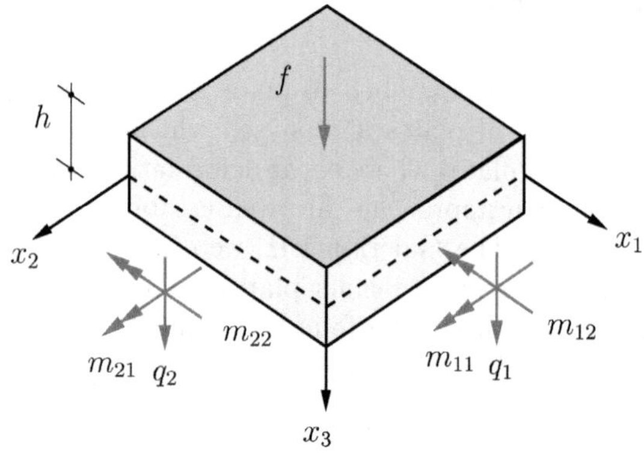

Figure 7.1: *Plate forces and moments*

The differential equation for the bending of the isotropic Kirchhoff plate is

$$D\,\Delta\Delta w = f \qquad \overset{\mathcal{F}}{\leftrightarrow} \qquad D(\hat{x}_1^2 + \hat{x}_2^2)^2\hat{w} = \hat{f}, \tag{7.3}$$

with f as transversal load per unit area. Thus the differential operator is

$$\mathcal{P}(\partial) = D\Delta\Delta \qquad \overset{\mathcal{F}}{\leftrightarrow} \qquad \hat{\mathcal{P}}(\hat{x}) = D(\hat{x}_1^2 + \hat{x}_2^2)^2. \tag{7.4}$$

The inverse of $\hat{\mathcal{P}}(\hat{x})$ is the transformed fundamental solution

$$W(x) = \frac{|x|^2}{8\pi D}(\ln|x| - \ln|r_0|) \qquad \overset{\mathcal{F}}{\leftrightarrow} \qquad \hat{W}(\hat{x}) = \frac{1}{D(\hat{x}_1^2 + \hat{x}_2^2)^2}, \tag{7.5}$$

with $|x| = \sqrt{x_1^2 + x_2^2}$, $|\hat{x}| = \sqrt{\hat{x}_1^2 + \hat{x}_2^2}$. For the original space, the fundamental solution is taken from the literature, e.g. [FB98]. The constant r_0 is arbitrary and often chosen such that $\ln r_0^2 = 1$ or just as $r_0 = 1$[2].

The boundary differential operators

The boundary quantities which are relevant in the context of BEM are the deflection w, the normal slope φ_ν, the normal bending moment m_ν, and the equivalent Kirchhoff shear $q_\nu = \nu_k q_k + \mathrm{d}m_T/\mathrm{d}s$ (we assume that the normal vector ν is piecewise constant)

$$\begin{array}{lllllll} w & & & \overset{\mathcal{F}}{\leftrightarrow} & \hat{w}; & & \\ \varphi_\nu & = & \nu_k \partial_k w & \overset{\mathcal{F}}{\leftrightarrow} & \hat{\varphi}_\nu & = & i\nu_k \hat{x}_k \hat{w}; \\ m_\nu & = & \nu_k \nu_l m_{kl} & \overset{\mathcal{F}}{\leftrightarrow} & \hat{m}_\nu & = & \nu_k \nu_l \hat{m}_{kl}; \\ q_\nu & = & \nu_k q_k + \tau_k \partial_k m_{lm} \nu_l \tau_m & \overset{\mathcal{F}}{\leftrightarrow} & \hat{q}_\nu & = & \nu_k \hat{q}_k + i\tau_k \hat{x}_k \hat{m}_{lm} \nu_l \tau_m, \end{array} \tag{7.6}$$

where the twisting moment for the Kirchhoff shear is defined as $m_\mathrm{T} = m_{kl}\nu_k\tau_l$, and the differentiation with respect to the arc length s of the boundary is $\mathrm{d}/\mathrm{d}s = \tau_k \partial_k$. At the i-th corner, there is a particular corner force f_c^i which is related to a jump of the twisting moment m_T around the corner

$$f_\mathrm{c}^i(x = x_\mathrm{c}^i) \quad = \quad m_\mathrm{T}^+(x = x_\mathrm{c}^i) - m_\mathrm{T}^-(x = x_\mathrm{c}^i). \tag{7.7}$$

This corner force is for arbitrary angles[3]

$$f_\mathrm{c}^i = \mathcal{A}_\mathrm{c}^i w \qquad \overset{\mathcal{F}}{\leftrightarrow} \qquad \hat{f}_\mathrm{c}^i = \hat{\mathcal{A}}_\mathrm{c}^i \hat{w}, \tag{7.8}$$

[2]The mathematical meaning of this arbitrariness is that one can add arbitrary homogeneous solutions to the fundamental solution. For all r_0 we have

$$D\,\Delta\Delta \left\{ \frac{1}{8\pi D} |x|^2 \ln |r_0| \right\} = 0.$$

The fundamental solution is the convolutional inverse of the differential operator. This is independent of the choice of r_0. The following is valid for all r_0

$$\mathcal{P}(\partial) W = \delta \qquad \overset{\mathcal{F}}{\leftrightarrow} \qquad \hat{\mathcal{P}}(\hat{x}) \hat{W} = 1.$$

In the transformed space, this arbitrariness lies in the freedom to choose a particular regularization procedure for the singularity at $|\hat{x}| = 0$, cf. [Dud97b].

[3]These particular corner terms can also be derived by a rigorous distributional reasoning as discussed in Chapter 3.2.

with the differential operator (see Fig.7.2 for the definition of ν^{+i}, ν^{-i})

$$\begin{aligned}
\mathcal{A}^i_{\mathrm{c}} &= (1-\bar{\nu})D(\nu_1^{+i}\nu_2^{+i} - \nu_1^{-i}\nu_2^{-i})(\partial_{11} - \partial_{22}) \\
&+ 2D(\nu_1^{+i}\nu_1^{+i} - \nu_2^{+i}\nu_2^{+i} - \nu_1^{-i}\nu_1^{-i} + \nu_2^{-i}\nu_2^{-i})\partial_{12}. \\
\stackrel{\mathcal{F}}{\leftrightarrow} \quad \hat{\mathcal{A}}^i_{\mathrm{c}} &= -(1-\bar{\nu})D(\nu_1^{+i}\nu_2^{+i} - \nu_1^{-i}\nu_2^{-i})(\hat{x}_1^2 - \hat{x}_2^2) \\
&+ 2D(\nu_1^{+i}\nu_1^{+i} - \nu_2^{+i}\nu_2^{+i} - \nu_1^{-i}\nu_1^{-i} + \nu_2^{-i}\nu_2^{-i})\hat{x}_1\hat{x}_2.
\end{aligned}$$

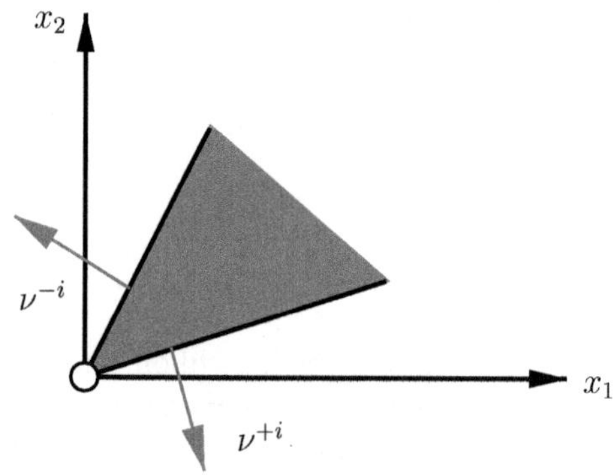

Figure 7.2: Definition of the normal vectors ν^{+i}, ν^{-i} for the corner term

Thus, the three boundary differential operators $\mathcal{A}^i_{\phi,\mathrm{m},\mathrm{q}}$ for a boundary element with the normal ν^i and the corner differential operator $\mathcal{A}^i_{\mathrm{c}}$ are defined as

$$\begin{aligned}
&\text{for } \varphi_\nu: && \mathcal{A}^i_\varphi = \nu^i_k \partial_k && \stackrel{\mathcal{F}}{\leftrightarrow} \; \hat{\mathcal{A}}^i_\varphi = i\nu_k \hat{x}_k; \\
&\text{for } m_\nu: && \mathcal{A}^i_{\mathrm{m}} = -K_{klmn}\nu^i_k\nu^i_l\partial_{mn} && \stackrel{\mathcal{F}}{\leftrightarrow} \; \hat{\mathcal{A}}^i_{\mathrm{m}} = K_{klmn}\nu^i_k\nu^i_l\hat{x}_m\hat{x}_n; \\
&\text{for } q_\nu: && \mathcal{A}^i_{\mathrm{q}} = -D\nu^i_k\partial_{kll} - && \stackrel{\mathcal{F}}{\leftrightarrow} \; \hat{\mathcal{A}}^i_{\mathrm{q}} = iD\nu^i_k\hat{x}_k\hat{x}_l\hat{x}_l + \\
& && \quad -K_{klmn}\tau^i_p\nu^i_k\tau^i_l\partial_{mnp} && \quad +iK_{klmn}\tau^i_p\nu^i_k\tau^i_l\hat{x}_m\hat{x}_n\hat{x}_p; \\
&\text{for } f_{\mathrm{c}}: && \mathcal{A}^i_{\mathrm{c}} \;\; \text{see (7.8)} && \stackrel{\mathcal{F}}{\leftrightarrow} \; \hat{\mathcal{A}}^i_{\mathrm{c}} \;\; \text{see (7.8).}
\end{aligned} \tag{7.9}$$

The boundary quantity and its transform are obtained via $\mathcal{A}^i_{\mathrm{k}}w$ or $\hat{\mathcal{A}}^i_{\mathrm{k}}\hat{w}$ $(k = \varphi, m, q, c)$.

For any well-posed problem, half of the boundary data must be giv en.For each pair of dual v ariables (dual in the sense that the pairs lead to work terms) either one of these two variables or the relation between both (a Robin type boundary condition) must be prescribed. There are four different types of boundary condition :

$$\begin{array}{llll} w(x) = w_\Gamma(x) & \text{for} \quad x \in \Gamma_\mathrm{w}; & \varphi_\nu(x) = \varphi_{\nu\Gamma}(x) & \text{for} \quad x \in \Gamma_\varphi; \\ m_\nu(x) = m_{\nu\Gamma}(x) & \text{for} \quad x \in \Gamma_\mathrm{m}; & q_\nu(x) = q_{\nu\Gamma}(x) & \text{for} \quad x \in \Gamma_\mathrm{q}. \end{array} \tag{7.10}$$

In addition, we have to prescribe for each corner point x^i_c either the jump of the twisting moment $f^i_\mathrm{c}(x^i_\mathrm{c})$ or the displacement $w(x^i_\mathrm{c})$ at this point.

For establishing the BIE, the following derivatives of the fundamental solution W are required : The fundamental slope Φ_ν (with $\ln r_0^2 = 1$)

$$\Phi_\nu = \mathcal{A}^i_\varphi W = \frac{\nu^i_k x_k}{4\pi D} \ln|x| \qquad \overset{\mathcal{F}}{\leftrightarrow} \qquad \hat{\Phi}_\nu = \hat{\mathcal{A}}^i_\varphi \hat{W} = \frac{i\nu_k \hat{x}_k}{D(\hat{x}_1^2 + \hat{x}_2^2)^2}, \tag{7.11}$$

the fundamental normal moment, cf. [Bes91] for the original space,

$$\begin{aligned} M_\nu &= \mathcal{A}^i_\mathrm{m} W = -\frac{1}{8\pi}\left[2(1+\bar{\nu})\ln|x| + (3+\bar{\nu})(\nu^i \cdot \nabla|x|)^2 \right. \\ &\qquad\qquad \left. +(1+3\bar{\nu})(\tau^i \cdot \nabla|x|)^2\right] \\ \overset{\mathcal{F}}{\leftrightarrow} \quad \hat{M}_\nu &= \hat{\mathcal{A}}^i_\mathrm{m} \hat{W} = \frac{K_{klmn}\nu^i_k \nu^i_l \hat{x}_m \hat{x}_n}{D(\hat{x}_1^2 + \hat{x}_2^2)^2}, \end{aligned} \tag{7.12}$$

and the fundamental Kirchhoff shear

$$\begin{aligned} Q_\nu &= \mathcal{A}^i_\mathrm{q} W = -\frac{1}{4\pi|x|}\left[2\nu^i \cdot \nabla|x| \right. \\ &\qquad \left. +(1-\bar{\nu})(\nu^i \cdot \nabla|x| - \varkappa|x|)((\nu^i \cdot \nabla|x|)^2 - (\tau^i \cdot \nabla|x|)^2)\right] \\ \overset{\mathcal{F}}{\leftrightarrow} \quad \hat{Q}_\nu &= \hat{\mathcal{A}}^i_\mathrm{q} \hat{W} = \frac{i\nu^i_k \hat{x}_k \hat{x}_l \hat{x}_l}{(\hat{x}_1^2 + \hat{x}_2^2)^2} + \frac{K_{klmn}\tau^i_p \nu^i_k \tau^i_l \hat{x}_m \hat{x}_n \hat{x}_p}{D(\hat{x}_1^2 + \hat{x}_2^2)^2}. \end{aligned} \tag{7.13}$$

The fundamental corner force is

$$F_\mathrm{c} = \mathcal{A}^i_\mathrm{c} W \qquad \overset{\mathcal{F}}{\leftrightarrow} \qquad \hat{F}_\mathrm{c} = \hat{\mathcal{A}}^i_\mathrm{c} \hat{W}, \tag{7.14}$$

The evaluation of the derivatives of the fundamental solution in Fourier space is very easy compared to that in the original space.

The symmetric Galerkin BIE

The Somigliana identity in analogy to (2.7) as a weak form equivalent to

(7.3) is, cf. [FB98, Jah98],

$$\begin{aligned}\kappa(x)w(x) &= \int_\Omega f(y)W(x-y)\,\mathrm{d}\Omega \\ &+ \int_\Gamma q_\nu(y)W(x-y)\,\mathrm{d}\Gamma_y - \int_\Gamma m_\nu(y)\Phi_\nu(x-y)\,\mathrm{d}\Gamma_y \\ &+ \int_\Gamma \varphi_\nu(y)M_\nu(x-y)\,\mathrm{d}\Gamma_y - \int_\Gamma w(y)Q_\nu(x-y)\,\mathrm{d}\Gamma_y \\ &+ \sum_i f_\mathrm{c}(y_\mathrm{c}^i)W(x-y_\mathrm{c}^i) - \sum_i w(y_\mathrm{c}^i)F_\mathrm{c}(x-y_\mathrm{c}^i).\end{aligned} \tag{7.15}$$

According to [Bes91], the free term is $\kappa = \Delta\psi/(2\pi)$, i.e. the percentage of the total angle 2π. The Galerkin version is obtained by additional weighting with test function ϕ_q^j (q is the dual variable of w). It is in distributional notation

$$\begin{aligned}\langle\phi_\mathrm{q}^j, w_\chi\rangle &= \langle\phi_\mathrm{q}^j, f * W\rangle + \langle\phi_\mathrm{q}^j, q_\nu * W\rangle - \langle\phi_\mathrm{q}^j, m_\nu * \Phi_\nu\rangle \\ &+ \langle\phi_\mathrm{q}^j, \varphi_\nu * M_\nu\rangle - \langle\phi_\mathrm{q}^j, w * Q_\nu\rangle \\ &+ \sum_i \langle\phi_\mathrm{q}^j, f_\mathrm{c}(y_\mathrm{c}^i)W(x-y_\mathrm{c}^i)\rangle - \sum_i \langle\phi_\mathrm{q}^j, w(y_\mathrm{c}^i)F_\mathrm{c}(x-y_\mathrm{c}^i)\rangle.\end{aligned} \tag{7.16}$$

Once more, the boundary factor κ is obtained implicitely as shown in the paradigmatic example for the heat equation by $w_\chi = \chi(x)w(x)$ as defined in (3.4). We introduce now the discretizations for all boundary quantities. Because of the high order of the differential operator (order four) we have to respect certain continuity requirements for the trial functions. The approximation of the deflection w must be C^1 at the nodes (the tangential derivative must be continuous) and the trial functions for the slope should be C^0, i.e. continuous. Therefore, the boundary values are approximated by Hermite polynomials [FB98], $(.)' = \mathrm{d}(.)/\,\mathrm{d}s = \tau_k^i\partial_k(.)$ is the tangential derivative,

$$w(x) \approx \sum_i \mathfrak{w}^i\phi_\mathrm{w}^i(x) + \mathfrak{p}^i\phi_\varphi^i(x). \tag{7.17}$$

The functions $\phi_\mathrm{w}^i, \phi_\varphi^i$ are constructed from the trial functions for the reference element (L_e^i is the length of the i-th element)

$$\phi_\mathrm{w}^0 = \begin{cases} x_1^2(3-2x_1) & \text{for the } (i-1)\text{-th element} \\ (x_1-1)^2(1+2x_1) & \text{for the } i\text{-th element} \\ 0 & \text{otherwise} \end{cases}$$
$$\phi_\varphi^0 = \begin{cases} x_1^2(x_1-1)L_\mathrm{e}^{i-1} & \text{for the } (i-1)\text{-th element} \\ x_1(x_1-1)^2L_\mathrm{e}^i & \text{for the } i\text{-th element} \\ 0 & \text{otherwise} \end{cases} \tag{7.18}$$

These functions have the interpolation properties

$$\phi^0_{\rm w}(x = x^l) = \delta_{kl} \qquad \frac{\rm d}{{\rm d}s}\phi^0_{\rm w}(x = x^l) = 0 \tag{7.19}$$
$$\phi^0_{\varphi}(x = x^l) = 0 \qquad \frac{\rm d}{{\rm d}s}\phi^0_{\varphi}(x = x^l) = \delta_{kl}.$$

The other quantities of the boundary, i.e. m_ν and q_ν, are approximated by piecewise linear polynomials

$$\phi^0_{\rm m} = \phi^0_{\rm q} = \begin{cases} x_1 & \text{for the } (i-1)\text{-th element} \\ 1 - x_1 & \text{for the } i\text{-th element} \\ 0 & \text{otherwise} \end{cases} \tag{7.20}$$

These trial functions are in distributional notation:

$$\begin{aligned} \phi^0_{\rm w} &= x_1^2(3 - 2x_1){\rm H}(x_1){\rm H}(1 - x_1)\delta(x_2) \\ \overset{\mathcal{F}}{\leftrightarrow} \quad \hat{\phi}^0_{\rm w} &= \frac{-12 + 6i\hat{x}_1 + i\hat{x}_1^3{\rm e}^{-i\hat{x}_1} + 6i\hat{x}_1{\rm e}^{-i\hat{x}_1} + 12{\rm e}^{-i\hat{x}_1}}{\hat{x}_1^4}; \\ \phi^0_{\rm w} &= (x_1 - 1)^2(1 + 2x_1){\rm H}(x_1){\rm H}(1 - x_1)\delta(x_2) \\ \overset{\mathcal{F}}{\leftrightarrow} \quad \hat{\phi}^0_{\rm w} &= \frac{12 - i\hat{x}_1^3 - 6i\hat{x}_1 - 6i\hat{x}_1{\rm e}^{-i\hat{x}_1} - 12\hat{x}_1{\rm e}^{-i\hat{x}_1}}{\hat{x}_1^4}. \end{aligned} \tag{7.21}$$

For the slope trial functions we get

$$\begin{aligned} \phi^0_{\varphi} &= x_1^2(x_1 - 1)l_{j-1}{\rm H}(x_1){\rm H}(1 - x_1)\delta(x_2) \\ \overset{\mathcal{F}}{\leftrightarrow} \quad \hat{\phi}^0_{\varphi} &= L_{\rm e}^{i-1}\frac{6 + \hat{x}_1^2{\rm e}^{-i\hat{x}_1} - 4i\hat{x}_1{\rm e}^{-i\hat{x}_1} - 6{\rm e}^{-i\hat{x}_1} - 2i\hat{x}_1}{\hat{x}_1^4}; \\ \phi^0_{\varphi} &= x_1(x_1 - 1)^2 l_j{\rm H}(x_1){\rm H}(1 - x_1)\delta(x_2) \\ \overset{\mathcal{F}}{\leftrightarrow} \quad \hat{\phi}^0_{\varphi} &= l_j\frac{6 - \hat{x}_1^2 - 4i\hat{x}_1 - 2i\hat{x}_1{\rm e}^{-i\hat{x}_1} - 6{\rm e}^{-i\hat{x}_1}}{\hat{x}_1^4}. \end{aligned} \tag{7.22}$$

The transformations of the linear trial functions are

$$\begin{aligned} \phi^0_{\rm m,q} &= x_1{\rm H}(x_1){\rm H}(1 - x_1)\delta(x_2) \\ \overset{\mathcal{F}}{\leftrightarrow} \quad \hat{\phi}^0_{\rm m,q} &= \frac{-1 + 1\hat{x}_1{\rm e}^{-i\hat{x}_1} + {\rm e}^{-i\hat{x}_1}}{\hat{x}_1^2} \\ \phi^0_{\rm m,q} &= (1 - x_1){\rm H}(x_1){\rm H}(1 - x_1)\delta(x_2) \\ \overset{\mathcal{F}}{\leftrightarrow} \quad \hat{\phi}^0_{\rm m,q} &= \frac{1 - {\rm e}^{i\hat{x}_1} - i\hat{x}_1}{\hat{x}_1^2}. \end{aligned} \tag{7.23}$$

With the exception of the deflection itself, all boundary quantities are dependent on the normal vector, they are discontinuous at corner points. Hence we define two different nodal values at these corners.

The treatment of the corner force f_c and the corner deflection w_c in Fourier space is enabled by defining particular corner trial functions $\phi_c^i = \delta(x - x_c^i)$

$$\begin{aligned} f_c &= \mathfrak{f}^i\,\delta(x - x_c^i) &\overset{\mathcal{F}}{\leftrightarrow}\quad \hat{f}_c &= \mathfrak{f}^i\,\mathrm{e}^{-i<x_c^i,\hat{x}>}; \\ w_c &= \mathfrak{w}^i\,\delta(x - x_c^i) &\overset{\mathcal{F}}{\leftrightarrow}\quad \hat{w}_c &= \mathfrak{w}^i\,\mathrm{e}^{-i<x_c^i,\hat{x}>}. \end{aligned} \tag{7.24}$$

The contributions of the corner terms to (7.16) can be written as

$$\sum_i \left\langle \phi_q^j, f_c(y_c^i) W(x - y_c^i) \right\rangle = \sum_i \mathfrak{f}^i \left\langle \phi_q^j, \phi_c^i * W \right\rangle \tag{7.25}$$

$$\overset{\mathcal{F}}{\leftrightarrow}\quad \frac{1}{(2\pi)^2} \quad \sum_i \mathfrak{f}^i \left\langle \hat{\phi}_q^j(-\hat{x}), \hat{\phi}_c^i(\hat{x}) \hat{W}(\hat{x}) \right\rangle.$$

These discretizations result finally in the discretized Galerkin BIE

$$\begin{aligned} \left\langle \phi_q^j, w_\chi \right\rangle &= \left\langle \phi_q^j, f_\chi * W \right\rangle + \sum_i^{N_q} \mathfrak{q}^i \left\langle \phi_q^j, \phi_q^i * W \right\rangle - \sum_i^{N_m} \mathfrak{m}^i \left\langle \phi_q^j, \phi_m^i * \Phi_\nu \right\rangle \\ &+ \sum_i^{N_\varphi} \mathfrak{p}^i \left\langle \phi_q^j, \phi_\varphi^i * M_\nu \right\rangle - \sum_i^{N_w} \mathfrak{w}^i \left\langle \phi_q^j, \phi_w^i * Q_\nu \right\rangle \\ &+ \sum_i^{N_c} \mathfrak{f}^i \left\langle \phi_q^j, \phi_c^i * W \right\rangle - \sum_i^{N_c} \mathfrak{w}^i \left\langle \phi_q^j, \phi_c^i * F_c \right\rangle. \end{aligned} \tag{7.26}$$

with $\left\langle \phi_q^j, w_\chi \right\rangle = \left\langle \phi_q^j, \chi p_q^j \right\rangle$; p_q^j is the polynomial defined for the test function ϕ_q^j. The kernels of this BIE can be highly singular, cf. [FB98] for the regularization. A more formal presentation of this Galerkin BIE is

$$\begin{aligned} \left\langle \phi_q^j, w_\chi \right\rangle &= \left\langle \phi_q^j, f_\chi * W \right\rangle + \sum_i^{N_q} \mathfrak{q}^i \left\langle \phi_q^j, \phi_q^i * W \right\rangle - \sum_i^{N_m} \mathfrak{m}^i \left\langle \phi_q^j, \phi_m^i * \mathcal{A}_\varphi^i W \right\rangle \\ &+ \sum_i^{N_\varphi} \mathfrak{p}^i \left\langle \phi_q^j, \phi_\varphi^i * \mathcal{A}_m^i W \right\rangle - \sum_i^{N_w} \mathfrak{w}^i \left\langle \phi_q^j, \phi_w^i * \mathcal{A}_q^i W \right\rangle \\ &+ \sum_i^{N_c} \mathfrak{f}^i \left\langle \phi_q^j, \phi_c^i * W \right\rangle - \sum_i^{N_c} \mathfrak{w}^i \left\langle \phi_q^j, \phi_c^i * \mathcal{A}_c^i W \right\rangle, \end{aligned} \tag{7.27}$$

which finds its Fourier equivalent in (after cancelling $(2\pi)^{-2}$)

$$
\begin{aligned}
\left\langle \hat{\phi}^j_{\mathrm{q}}(-\hat{x}), \hat{w}_\chi \right\rangle &= \left\langle \hat{\phi}^j_{\mathrm{q}}(-\hat{x}), \hat{f}_\chi \hat{W} \right\rangle \\
&+ \sum_i^{N_{\mathrm{q}}} \mathfrak{q}^i \left\langle \hat{\phi}^j_{\mathrm{q}}(-\hat{x}), \hat{\phi}^i_{\mathrm{q}} \hat{W} \right\rangle - \sum_i^{N_{\mathrm{m}}} \mathfrak{m}^i \left\langle \hat{\phi}^j_{\mathrm{q}}(-\hat{x}), \hat{\phi}^i_{\mathrm{m}} \hat{\mathcal{A}}^i_\varphi \hat{W} \right\rangle \\
&+ \sum_i^{N_\varphi} \mathfrak{p}^i \left\langle \hat{\phi}^j_{\mathrm{q}}(-\hat{x}), \hat{\phi}^i_\varphi \hat{\mathcal{A}}^i_{\mathrm{m}} \hat{W} \right\rangle - \sum_i^{N_{\mathrm{w}}} \mathfrak{w}^i \left\langle \hat{\phi}^j_{\mathrm{q}}(-\hat{x}), \hat{\phi}^i_{\mathrm{w}} \hat{\mathcal{A}}^i_{\mathrm{q}} \hat{W} \right\rangle \\
&+ \sum_i^{N_{\mathrm{c}}} \mathfrak{f}^i \left\langle \hat{\phi}^j_{\mathrm{q}}(-\hat{x}), \hat{\phi}^i_{\mathrm{c}} \hat{W} \right\rangle - \sum_i^{N_{\mathrm{c}}} \mathfrak{w}^i \left\langle \hat{\phi}^j_{\mathrm{q}}(-\hat{x}), \hat{\phi}^i_{\mathrm{c}} \hat{\mathcal{A}}^i_{\mathrm{c}} \hat{W} \right\rangle .
\end{aligned} \tag{7.28}
$$

For a symmetric Galerkin method, additional BIEs are required which will be given in the following for Fourier space. If needed they can be transferred easily to the original space. The BIE for the normal slope is obtained by applying the adjoint of $\hat{\mathcal{A}}^j_\varphi = -\hat{\mathcal{A}}^j_\varphi$ on (7.28) and by choosing the dual test function. Here we need a moment test function ϕ^j_{m}. The BIE is then

$$
\begin{aligned}
-\left\langle \hat{\phi}^j_{\mathrm{m}}(-\hat{x}), \hat{\mathcal{A}}^j_\varphi \hat{w}_\chi \right\rangle &= -\left\langle \hat{\phi}^j_{\mathrm{m}}(-\hat{x}), \hat{f}_\chi \hat{\mathcal{A}}^j_\varphi \hat{W} \right\rangle \\
&- \sum_i^{N_{\mathrm{q}}} \mathfrak{q}^i \left\langle \hat{\phi}^j_{\mathrm{m}}(-\hat{x}), \hat{\phi}^i_{\mathrm{q}} \hat{\mathcal{A}}^j_\varphi \hat{W} \right\rangle + \sum_i^{N_{\mathrm{m}}} \mathfrak{m}^i \left\langle \hat{\phi}^j_{\mathrm{m}}(-\hat{x}), \hat{\phi}^i_{\mathrm{m}} \hat{\mathcal{A}}^j_\varphi \hat{\mathcal{A}}^i_\varphi \hat{W} \right\rangle \\
&- \sum_i^{N_\varphi} \mathfrak{p}^i \left\langle \hat{\phi}^j_{\mathrm{m}}(-\hat{x}), \hat{\phi}^i_\varphi \hat{\mathcal{A}}^j_\varphi \hat{\mathcal{A}}^i_{\mathrm{m}} \hat{W} \right\rangle + \sum_i^{N_{\mathrm{w}}} \mathfrak{w}^i \left\langle \hat{\phi}^j_{\mathrm{m}}(-\hat{x}), \hat{\phi}^i_{\mathrm{w}} \hat{\mathcal{A}}^j_\varphi \hat{\mathcal{A}}^i_{\mathrm{q}} \hat{W} \right\rangle \\
&- \sum_i^{N_{\mathrm{c}}} \mathfrak{f}^i \left\langle \hat{\phi}^j_{\mathrm{m}}(-\hat{x}), \hat{\phi}^i_{\mathrm{c}} \hat{\mathcal{A}}^j_\varphi \hat{W} \right\rangle + \sum_i^{N_{\mathrm{c}}} \mathfrak{w}^i \left\langle \hat{\phi}^j_{\mathrm{m}}(-\hat{x}), \hat{\phi}^i_{\mathrm{c}} \hat{\mathcal{A}}^j_\varphi \hat{\mathcal{A}}^i_{\mathrm{c}} \hat{W} \right\rangle .
\end{aligned} \tag{7.29}
$$

The other BIE are obtained analoguously. By using the operator notation introduced in (7.9), the following scheme can be established for the total system

$$
\left\langle \hat{\phi}^j(-\hat{x}), \hat{\mathcal{B}} \hat{w}_\chi \right\rangle = \left\langle \hat{\phi}^j(-\hat{x}), \hat{f}_\chi \hat{\mathcal{B}} \hat{W} \right\rangle + \sum_i \hat{\mathfrak{u}}^i \left\langle \hat{\phi}^j(-\hat{x}), \hat{\phi}^i \hat{\mathcal{A}} \hat{W} \right\rangle, \tag{7.30}
$$

with

$$
\begin{aligned}
\hat{\phi}^j &= \left(\hat{\phi}^j_{\mathrm{q}},\ \hat{\phi}^j_{\mathrm{m}},\ \hat{\phi}^j_\varphi,\ \hat{\phi}^j_{\mathrm{w}},\ \hat{\phi}^j_{\mathrm{c}},\ \hat{\phi}^j_{\mathrm{c}} \right) \\
\hat{\phi}^i &= \left(\textstyle\sum_i^{N_{\mathrm{q}}} \hat{\phi}^i_{\mathrm{q}},\ \sum_i^{N_{\mathrm{m}}} \hat{\phi}^i_{\mathrm{m}},\ \sum_i^{N_\varphi} \hat{\phi}^i_\varphi,\ \sum_i^{N_{\mathrm{w}}} \hat{\phi}^i_{\mathrm{w}},\ \sum_i^{N_{\mathrm{c}}} \hat{\phi}^i_{\mathrm{c}},\ \sum_i^{N_{\mathrm{c}}} \hat{\phi}^i_{\mathrm{c}} \right) \\
\hat{\mathfrak{u}}^i &= \left(\hat{\mathfrak{q}}^i, \hat{\mathfrak{m}}^i, \hat{\mathfrak{p}}^i, \hat{\mathfrak{w}}^i, \hat{\mathfrak{f}}^i, \hat{\mathfrak{w}}^i \right),
\end{aligned}
$$

and

$$\hat{\mathcal{A}} = \begin{bmatrix} \mathcal{I} & -\hat{\mathcal{A}}^i_\varphi & \hat{\mathcal{A}}^i_\mathrm{m} & -\hat{\mathcal{A}}^i_\mathrm{q} & \mathcal{I} & -\hat{\mathcal{A}}^i_\mathrm{c} \\ -\hat{\mathcal{A}}^j_\varphi & \hat{\mathcal{A}}^j_\varphi\hat{\mathcal{A}}^i_\varphi & -\hat{\mathcal{A}}^j_\varphi\hat{\mathcal{A}}^i_\mathrm{m} & \hat{\mathcal{A}}^j_\varphi\hat{\mathcal{A}}^i_\mathrm{q} & -\hat{\mathcal{A}}^j_\varphi & \hat{\mathcal{A}}^j_\varphi\hat{\mathcal{A}}^i_\mathrm{c} \\ \hat{\mathcal{A}}^j_\mathrm{m} & -\hat{\mathcal{A}}^j_\mathrm{m}\hat{\mathcal{A}}^i_\varphi & \hat{\mathcal{A}}^j_\mathrm{m}\hat{\mathcal{A}}^i_\mathrm{m} & -\hat{\mathcal{A}}^j_\mathrm{m}\hat{\mathcal{A}}^i_\mathrm{q} & \hat{\mathcal{A}}^j_\mathrm{m} & -\hat{\mathcal{A}}^j_\mathrm{m}\hat{\mathcal{A}}^i_\mathrm{c} \\ -\hat{\mathcal{A}}^j_\mathrm{q} & \hat{\mathcal{A}}^j_\mathrm{q}\hat{\mathcal{A}}^i_\varphi & -\hat{\mathcal{A}}^j_\mathrm{q}\hat{\mathcal{A}}^i_\mathrm{m} & \hat{\mathcal{A}}^j_\mathrm{q}\hat{\mathcal{A}}^i_\mathrm{q} & -\hat{\mathcal{A}}^j_\mathrm{q} & \hat{\mathcal{A}}^j_\mathrm{q}\hat{\mathcal{A}}^i_\mathrm{c} \\ \mathcal{I} & -\hat{\mathcal{A}}^i_\varphi & \hat{\mathcal{A}}^i_\mathrm{m} & -\hat{\mathcal{A}}^i_\mathrm{q} & \mathcal{I} & -\hat{\mathcal{A}}^i_\mathrm{c} \\ -\hat{\mathcal{A}}^j_\mathrm{c} & \hat{\mathcal{A}}^j_\mathrm{c}\hat{\mathcal{A}}^i_\varphi & -\hat{\mathcal{A}}^j_\mathrm{c}\hat{\mathcal{A}}^i_\mathrm{m} & \hat{\mathcal{A}}^j_\mathrm{c}\hat{\mathcal{A}}^i_\mathrm{q} & -\hat{\mathcal{A}}^j_\mathrm{c} & \hat{\mathcal{A}}^j_\mathrm{c}\hat{\mathcal{A}}^i_\mathrm{c} \end{bmatrix}$$

$$\mathcal{B} = \left(\mathcal{I} \; -\hat{\mathcal{A}}^j_\varphi \; \hat{\mathcal{A}}^j_\mathrm{m} \; -\hat{\mathcal{A}}^j_\mathrm{q} \; \mathcal{I} \; -\hat{\mathcal{A}}^j_\mathrm{c} \right).$$

$\mathcal{I}$ is the identity operator such that $\mathcal{I}\hat{W} = \hat{W}$. The transformed terms in the operator matrix $\hat{\mathcal{A}}$ of (7.30) are obtained by simple multiplication. The scalar product on the left-hand side of (7.29) can be used for regularization purposes in the original as well as in the transformed space. The differentiation $\mathcal{B}w_\chi = \mathcal{B}\{w(x)\chi(x)\}$ should be done carefully taking into account that the product of the deflection with the cutoff-distribution of the domain has to be evaluated. A detailed discussion of this regularization is not within the scope intended here. It will be published separately.

The free terms on the right-hand side

As an example, we investigate

$$\begin{aligned} \langle \phi^j_\mathrm{m}, \mathcal{A}^j_\varphi\{\chi w\}\rangle &= \langle \phi^j_\mathrm{m}, (\chi\nu^j_k\partial_k w + w\nu^j_k\partial_k\chi)\rangle \\ &= \langle \phi^j_\mathrm{m}, \chi\varphi_\nu\rangle + \langle \phi^j_\mathrm{m}, w\delta(\psi)\nu^j_k\partial_k\psi\rangle, \end{aligned} \tag{7.31}$$

with $\chi = \mathrm{H}(\psi)$ as cutoff-distribution of the domain Ω. The local support of the test function ϕ^j_m is part of the boundary, i.e. part of that which is described by $\delta(\psi)$. Therefore, we have for the last scalar product locally something of the character

$$\int_{\mathbb{R}^2} p(x)\delta(\psi)\delta(\psi)\,\mathrm{d}\Omega = C_1\int_{\mathbb{R}^2}\delta(\psi)^2\,\mathrm{d}\Omega = C_2\delta(0),$$

with some C^∞-function $p(x)$ and constants $C_{1,2}$. This apparently divergent integral is of character $\delta(0)$. The same is true for all differential operators $\mathcal{A}^j_{\phi,\mathrm{m,q,c}}$. According to the author's experience, all these singularities cancel with the singularities of the right-hand side. The proofs for this are beyond the scope of this thesis and will be published later.

Sample integrations / isotropic case

In section 5.1.2, we defined linear test and trial functions, for example

$$\phi_{\mathrm{q}}^1 = (1-x_1)\mathrm{H}(x_1)\mathrm{H}(1-x_1)\delta(x_2) \overset{\mathcal{F}}{\leftrightarrow} \hat{\phi}_{\mathrm{q}}^1 = \frac{1-\mathrm{e}^{-i\hat{x}_1}+i\hat{x}_1}{\hat{x}_1^2}. \tag{7.32}$$

An exemplary analytical integration of one entry of the matrix leads in the single layer case to (we have chosen $r_0 = \sqrt{\mathrm{e}}$)

$$\begin{aligned}
H^{11} &= \langle \phi_{\mathrm{q}}^1, \phi_{\mathrm{q}}^1 * W \rangle \\
&= \frac{1}{8D\pi}\int_0^1 (1-x_1)\int_0^1 (1-y_1)(x_1-y_1)^2 \ln\left(\sqrt{\frac{(x_1-y_1)^2}{\mathrm{e}}}\right) dy_1\, dx_1 \\
&= \frac{1}{D\pi}\int_0^1 \left[-\frac{13}{1152} - \frac{13x_1^2}{192} - \frac{x_1^3\ln|x_1-1|}{24} + \frac{\ln|x_1-1|^2}{192} + \frac{13x_1}{288}\right. \\
&\quad + \frac{x_1^4\ln|x_1-1|}{96} + \frac{x_1^3}{96} - \frac{x_1\ln|x_1-1|^2}{48} + \frac{x_1^2\ln|x_1-1|^2}{32} \\
&\quad \left. + \frac{x_1^3\ln|x_1|}{24} - \frac{x_1^4\ln|x_1|}{96}\right] dx_1 = -\frac{5}{1152D\pi}.
\end{aligned}$$

The corresponding integral in Fourier space is computed as

$$\begin{aligned}
H^{11} &= \frac{1}{(2\pi)^2}\left\langle \hat{\phi}_{\mathrm{q}}^1(-\hat{x}), \hat{\phi}_{\mathrm{q}}^1(\hat{x})\hat{W}(\hat{x})\right\rangle = \frac{1}{(2\pi)^2}\int_{\mathbb{R}^2} \hat{\phi}_{\mathrm{q}}^1(-\hat{x})\hat{\phi}_{\mathrm{q}}^1(\hat{x})\hat{W}(\hat{x})\, d\hat{x} \\
&= \frac{1}{D(2\pi)^2}\int_{\mathbb{R}^2} \frac{2-2\hat{x}_1\sin\hat{x}_1+\hat{x}_1^2-2\cos\hat{x}_1}{\hat{x}_1^4(\hat{x}_1^4+2\hat{x}_1^2\hat{x}_2^2+\hat{x}_2^4)}\, d\hat{x}_1\, d\hat{x}_2 \\
&= \frac{1}{2\pi D}\int_{\mathbb{R}^1}\left[\frac{1}{3\hat{x}_2^4} + \frac{5\mathrm{sgn}\hat{x}_2}{2\hat{x}_2^7} + \frac{5\mathrm{e}^{\hat{x}_2}}{4\hat{x}_2^7} - \frac{5\mathrm{e}^{-\hat{x}_2}}{4\hat{x}_2^7} + \frac{5\mathrm{sgn}\hat{x}_2\mathrm{e}^{\hat{x}_2}}{4\hat{x}_2^6} - \frac{5\mathrm{sgn}\hat{x}_2\mathrm{e}^{-\hat{x}_2}}{4\hat{x}_2^6}\right. \\
&\quad - \frac{5\mathrm{sgn}\hat{x}_2\mathrm{e}^{-\hat{x}_2}}{4\hat{x}_2^7} - \frac{5\mathrm{e}^{-\hat{x}_2}}{4\hat{x}_2^6} - \frac{5\mathrm{e}^{\hat{x}_2}}{4\hat{x}_2^6} + \frac{\mathrm{e}^{\hat{x}_2}}{4\hat{x}_2^5} - \frac{5\mathrm{sgn}\hat{x}_2\mathrm{e}^{\hat{x}_2}}{4\hat{x}_2^7} \\
&\quad \left. - \frac{\mathrm{sgn}\hat{x}_2\mathrm{e}^{-\hat{x}_2}}{4\hat{x}_2^5} - \frac{\mathrm{sgn}\hat{x}_2\mathrm{e}^{\hat{x}_2}}{4\hat{x}_2^5} - \frac{\mathrm{e}^{\hat{x}_2}}{4\hat{x}_2^5} - \frac{3\mathrm{sgn}\hat{x}_2}{4\hat{x}_2^5}\right] d\hat{x}_2 = -\frac{5}{1152D\pi}.
\end{aligned}$$

The kernel in Fourier space is hypersingular and is regularized as shown in the example for the isotropic heat equation.

Example : Fourier BEM for the clamped square plate

For a clamped square plate with $\Omega = [0,1]\times[0,1]$ and with two elements at each side, the system (7.30) can be reduced because of $w = 0, \varphi_\nu = 0$ on the total boundary and due to $w_{\mathrm{c}} = 0$ for all corner points. The right-hand side of the Fourier BIE is

$$\begin{pmatrix} \hat{\phi}^j_{\mathrm{q}}(-\hat{x}) \\ \hat{\phi}^j_{\mathrm{m}}(-\hat{x}) \\ \hat{\phi}^j_{\mathrm{c}}(-\hat{x}) \end{pmatrix}^T \begin{bmatrix} \mathcal{I} & -\hat{\mathcal{A}}^i_{\varphi} & \mathcal{I} \\ -\hat{\mathcal{A}}^j_{\varphi} & \hat{\mathcal{A}}^j_{\varphi}\hat{\mathcal{A}}^i_{\varphi} & -\hat{\mathcal{A}}^j_{\varphi} \\ \mathcal{I} & -\hat{\mathcal{A}}^i_{\varphi} & \mathcal{I} \end{bmatrix} \begin{pmatrix} \sum_i^{N_{\mathrm{q}}} \hat{\phi}^i_{\mathrm{q}} \hat{W} \\ \sum_i^{N_{\mathrm{m}}} \hat{\phi}^i_{\mathrm{m}} \hat{W} \\ \sum_i^{N_{\mathrm{c}}} \hat{\phi}^i_{\mathrm{c}} \hat{W} \end{pmatrix}. \tag{7.33}$$

For the volume forces f we get

$$\begin{pmatrix} \hat{\phi}^j_{\mathrm{q}}(-\hat{x}) \\ \hat{\phi}^j_{\mathrm{m}}(-\hat{x}) \\ \hat{\phi}^j_{\mathrm{c}}(-\hat{x}) \end{pmatrix}^T \begin{pmatrix} \mathcal{I} \\ -\hat{\mathcal{A}}^j_{\varphi} \\ \mathcal{I} \end{pmatrix} \hat{f}_{\chi} \hat{W}. \tag{7.34}$$

The free term on the left-hand side is zero because of $w = \varphi_{\nu} = 0$ along the boundary. Linear trial and test functions are chosen for the normal moment and the Kirchhoff shear forces at the boundary. For the corner forces we have the four trial functions

$$\begin{aligned} \phi^1_{\mathrm{c}} &= \delta(x_1)\delta(x_2) & &\overset{\mathcal{F}}{\leftrightarrow} & \hat{\phi}^2_{\mathrm{c}} &= 1; \\ \phi^2_{\mathrm{c}} &= \delta(x_1 - 1)\delta(x_2) & &\overset{\mathcal{F}}{\leftrightarrow} & \hat{\phi}^2_{\mathrm{c}} &= \mathrm{e}^{-i\hat{x}_1}; \\ \phi^3_{\mathrm{c}} &= \delta(x_1 - 1)\delta(x_2 - 1) & &\overset{\mathcal{F}}{\leftrightarrow} & \hat{\phi}^3_{\mathrm{c}} &= \mathrm{e}^{-i(\hat{x}_1 + \hat{x}_2)}; \\ \phi^4_{\mathrm{c}} &= \delta(x_1)\delta(x_2 - 1) & &\overset{\mathcal{F}}{\leftrightarrow} & \hat{\phi}^4_{\mathrm{c}} &= \mathrm{e}^{-i\hat{x}_2}. \end{aligned} \tag{7.35}$$

Therefore, we get the following system of 36 equations ($N_{\mathrm{q}} = 16, N_{\mathrm{m}} = 16, N_{\mathrm{c}} = 4$):

$$\begin{aligned} j = 1 \ldots N_{\mathrm{q}}\colon 0 &= \left\langle \hat{\phi}^j_{\mathrm{q}}(-\hat{x}), \hat{f}_{\chi} \hat{W} \right\rangle + \left\langle \hat{\phi}^j_{\mathrm{q}}(-\hat{x}), \sum_i^{N_{\mathrm{q}}} \hat{\phi}^i_{\mathrm{q}} \hat{W} \right\rangle \\ &- \left\langle \hat{\phi}^j_{\mathrm{q}}(-\hat{x}), \sum_i^{N_{\mathrm{m}}} \hat{\phi}^i_{\mathrm{m}} \hat{\mathcal{A}}^i_{\varphi} \hat{W} \right\rangle + \left\langle \hat{\phi}^j_{\mathrm{q}}(-\hat{x}), \sum_i^{N_{\mathrm{c}}} \hat{\phi}^i_{\mathrm{c}} \hat{W} \right\rangle \\ j = 1 \ldots N_{\mathrm{m}}\colon 0 &= -\left\langle \hat{\phi}^j_{\mathrm{m}}(-\hat{x}), \hat{f}_{\chi} \hat{\mathcal{A}}^j_{\varphi} \hat{W} \right\rangle - \left\langle \hat{\phi}^j_{\mathrm{m}}(-\hat{x}), \sum_i^{N_{\mathrm{q}}} \hat{\phi}^i_{\mathrm{q}} \hat{\mathcal{A}}^j_{\varphi} \hat{W} \right\rangle \\ &+ \left\langle \hat{\phi}^j_{\mathrm{m}}(-\hat{x}), \sum_i^{N_{\mathrm{m}}} \hat{\phi}^i_{\mathrm{m}} \hat{\mathcal{A}}^j_{\varphi} \hat{\mathcal{A}}^i_{\varphi} \hat{W} \right\rangle - \left\langle \hat{\phi}^j_{\mathrm{m}}(-\hat{x}), \sum_i^{N_{\mathrm{c}}} \hat{\phi}^i_{\mathrm{c}} \hat{\mathcal{A}}^j_{\varphi} \hat{W} \right\rangle \\ j = 1 \ldots N_{\mathrm{c}}\colon 0 &= -\left\langle \hat{\phi}^j_{\mathrm{c}}(-\hat{x}), \hat{f}_{\chi} \hat{W} \right\rangle - \left\langle \hat{\phi}^j_{\mathrm{c}}(-\hat{x}), \sum_i^{N_{\mathrm{q}}} \hat{\phi}^i_{\mathrm{q}} \hat{W} \right\rangle \\ &+ \left\langle \hat{\phi}^j_{\mathrm{c}}(-\hat{x}), \sum_i^{N_{\mathrm{m}}} \hat{\phi}^i_{\mathrm{m}} \hat{\mathcal{A}}^i_{\varphi} \hat{W} \right\rangle - \left\langle \hat{\phi}^j_{\mathrm{c}}(-\hat{x}), \sum_i^{N_{\mathrm{c}}} \hat{\phi}^i_{\mathrm{c}} \hat{W} \right\rangle. \end{aligned} \tag{7.36}$$

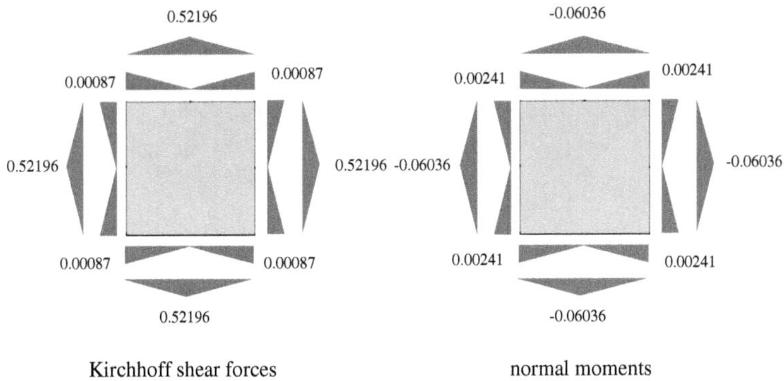

Figure 7.3: The equivalent Kirchhoff shear q_ν and the normal moments m_ν at the boundaries caused by uniform loading of the clamped thin plate (4×4 elements)

A uniform loading

$$\begin{aligned} f &= \mathrm{H}(x_1)\mathrm{H}(2-x_1)\mathrm{H}(x_2)\mathrm{H}(2-x_2) \\ \overset{\mathcal{F}}{\leftrightarrow} \quad \hat{f} &= -\frac{(1-\mathrm{e}^{-2i\hat{x}_1})(1-\mathrm{e}^{-2i\hat{x}_2})}{\hat{x}_1\hat{x}_2} \end{aligned} \tag{7.37}$$

was applied. The final results for the equivalent Kirchhoff forces and the normal moments at the boundaries are giv enin Fig.7.3. All integrations were computed analytically in F ourierspace. For the moment as well as for the shear force, the v alues at the corner are theoretical zero which is approximately fulfilled even b y this coarse mesh. The displacements w, the slope $\varphi_1 = \partial_1 w$, the moment m_{11}, and the shear force q_ν in the interior are shown in Fig.7.4.

7.1.2 Orthotropic case

The differential equation and its fundamental solution

We define the coordinates x_1, x_2 as the principal orientations of the material. The oriented flexural rigidities D_{11}, D_{12}, D_{22} are

$$D_{11} = \frac{E_1 h^3}{12(1-\bar{\nu}_{12}\bar{\nu}_{21})}, \quad D_{12} = \frac{G_{12}h^3}{12}, \quad D_{22} = \frac{E_2 h^3}{12(1-\bar{\nu}_{21}\bar{\nu}_{12})}. \tag{7.38}$$

with $E_k, k = 1, 2$ as the Young modulus in x_k-direction, G_{12} as the shear modulus, and $\bar{\nu}_{kl}$ as the corresponding P oisson'scoefficients. The moments

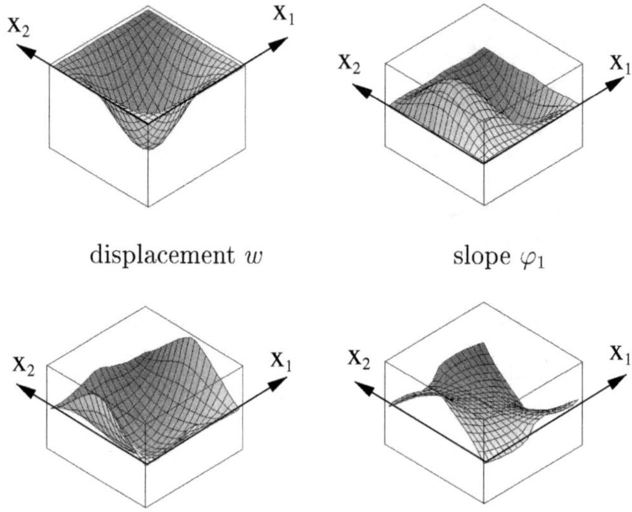

displacement w slope φ_1

moment m_{11} shear force q_1

Figure 7.4: The clamped thin plate under uniform loading (4 × 4 elements)

are

$$\begin{array}{rclcrcl} m_{11} & = & -D_{11}\left(\partial_{11}+\bar{\nu}_{12}\,\partial_{22}\right)w & \overset{\mathcal{F}}{\leftrightarrow} & \hat{m}_{11} & = & D_{11}\left(\hat{x}_1^2+\bar{\nu}_{12}\hat{x}_2^2\right)\hat{w}; \\ m_{12} & = & -2D_{12}\partial_{12}w & \overset{\mathcal{F}}{\leftrightarrow} & \hat{m}_{12} & = & 2D_{12}\hat{x}_1\hat{x}_2\hat{w}, \\ m_{22} & = & -D_{22}\left(\partial_{22}+\bar{\nu}_{21}\partial_{11}\right)w & \overset{\mathcal{F}}{\leftrightarrow} & \hat{m}_{22} & = & D_{22}\left(\hat{x}_2^2+\bar{\nu}_{21}\hat{x}_1^2\right)\hat{w}; \end{array} \tag{7.39}$$

and as shearing forces we get

$$\begin{array}{rclcrcl} q_1 & = & -\partial_1\left(D_{11}\partial_{11}+D_3\partial_{22}\right)w & \overset{\mathcal{F}}{\leftrightarrow} & \hat{q}_1 & = & i\hat{x}_1\left(D_{11}\hat{x}_1^2+D_3\hat{x}_2^2\right)\hat{w}; \\ q_2 & = & -\partial_2\left(D_{22}\partial_{22}+D_3\partial_{11}\right)w & \overset{\mathcal{F}}{\leftrightarrow} & \hat{q}_2 & = & i\hat{x}_2\left(D_{22}\hat{x}_2^2+D_3\hat{x}_1^2\right)\hat{w}, \end{array} \tag{7.40}$$

with $D_3 = \bar{\nu}_{21}D_{11} + 2D_{12} = \bar{\nu}_{12}D_{22} + 2D_{12}$. Hence, the differential equation for orthotropic plates is, cf. [Lek68],

$$\begin{array}{rcl} \left(D_{11}\partial_{1111}+2D_3\partial_{1122}+D_{22}\partial_{2222}\right)w & = & f \\ \overset{\mathcal{F}}{\leftrightarrow} \quad \left(D_{11}\hat{x}_1^4+2D_3\hat{x}_1^2\hat{x}_2^2+D_{22}\hat{x}_2^4\right)\hat{w} & = & \hat{f}. \end{array} \tag{7.41}$$

Which leads to the fundamental solution in Fourier space

$$\hat{W} = \frac{1}{D_{11}\hat{x}_1^4+2D_3\hat{x}_1^2\hat{x}_2^2+D_{22}\hat{x}_2^4}. \tag{7.42}$$

The symmetric Galerkin BIE for orthotropic plates

To establish the Fourier BIE, the following boundary differential operators are derived from (7.39) and (7.40) together with (7.6)

$$\begin{aligned}
\hat{\mathcal{A}}^i_\varphi &= i\nu^i_k \hat{x}_k; \qquad (7.43)\\
\hat{\mathcal{A}}^i_{\mathrm{m}} &= \nu^i_1\nu^i_1(\hat{x}^2_1 + \bar{\nu}_{12}\hat{x}^2_2)D_{11} + 4\nu^i_1\nu^i_2\hat{x}_1\hat{x}_2 D_{12} + \nu^i_2\nu^i_2(\hat{x}^2_2 + \bar{\nu}_{12}\hat{x}^2_1)D_{22};\\
\hat{\mathcal{A}}^i_{\mathrm{q}} &= i\nu^i_1(\hat{x}_1 + \nu^i_2\nu^i_2\hat{x}_1 - \nu^i_1\nu^i_2\hat{x}_2)(\hat{x}^2_1 + \bar{\nu}_{12}\hat{x}^2_2)D_{11}\\
&+ 4i(\nu^i_2\nu^i_2\nu^i_2\hat{x}_1 + \nu^i_1\nu^i_1\nu^i_1\hat{x}_2)\hat{x}_1\hat{x}_2 D_{12}\\
&+ i\nu^i_2(\hat{x}_2 + \nu^i_1\nu^i_1\hat{x}_2 - \nu^i_2\nu^i_1\hat{x}_1)(\hat{x}^2_2 + \bar{\nu}_{12}\hat{x}^2_1)D_{22};\\
\mathcal{A}^i_{\mathrm{c}} &= -(\nu^{+i}_1\nu^{+i}_2 - \nu^{-i}_1\nu^{-i}_2)\left[(\hat{x}^2_1 + \bar{\nu}_{12}\hat{x}^2_2)D_{11} - (\hat{x}^2_2 + \bar{\nu}_{12}\hat{x}^2_1)D_{22}\right]\\
&+ 2(\nu^{+i}_1\nu^{+i}_1 - \nu^{+i}_2\nu^{+i}_2 - \nu^{-i}_1\nu^{-i}_1 + \nu^{-i}_1\nu^{-i}_1)\hat{x}_1\hat{x}_2 D_{12}.
\end{aligned}$$

Introducing (7.42) and (7.43) in (7.30) leads to the Galerkin BIE for orthotropic thin plates.

7.1.3 General anisotropic case

The differential equation and its fundamental solution

The strain-stress relations for general anisotropic plates are with the flexibilities a_{kl} :

$$\begin{aligned}
\varepsilon_{11} &= a_{11}\sigma_{11} + a_{12}\sigma_{22} + a_{16}\sigma_{12};\\
\varepsilon_{22} &= a_{12}\sigma_{11} + a_{22}\sigma_{22} + a_{26}\sigma_{12}; \qquad (7.44)\\
\varepsilon_{12} &= a_{16}/2\sigma_{11} + a_{26}/2\sigma_{22} + a_{66}/2\sigma_{12}.
\end{aligned}$$

They are linked to the corresponding rigidities D_{kl} by

$$\begin{aligned}
D_{11} &= c_1(a_{22}a_{66} - a^2_{26}); & D_{22} &= c_1(a_{11}a_{66} - a^2_{16});\\
D_{12} &= c_1(a_{16}a_{26} - a_{12}a_{66}); & D_{66} &= c_1(a_{11}a_{22} - a^2_{12}); \qquad (7.45)\\
D_{16} &= c_1(a_{12}a_{26} - a_{22}a_{16}); & D_{26} &= c_1(a_{12}a_{16} - a_{11}a_{26});
\end{aligned}$$

with

$$\frac{1}{c_1} = \frac{12}{h^3}\det\begin{bmatrix} a_{11} & a_{12} & a_{16}\\ a_{12} & a_{22} & a_{26}\\ a_{16} & a_{26} & a_{66}\end{bmatrix}.$$

The bending and twisting moments are

$$\begin{aligned}
m_{11} &= -(D_{11}\partial_{11} + D_{12}\partial_{22} + 2D_{16}\partial_{12})w;\\
m_{12} &= -(D_{16}\partial_{11} + D_{26}\partial_{22} + 2D_{66}\partial_{12})w; \qquad (7.46)\\
m_{22} &= -(D_{12}\partial_{11} + D_{22}\partial_{22} + 2D_{26}\partial_{12})w.
\end{aligned}$$

and the shear forces are

$$\begin{array}{rcl} q_1 &=& -[D_{11}\partial_{111} + 3D_{16}\partial_{112} + (D_{12}+2D_{66})\partial_{122} + D_{26}\partial_{222}]; \\ q_2 &=& -[D_{16}\partial_{111} + (D_{12}+2D_{66})\partial_{112} + 3D_{26}\partial_{122} + D_{22}\partial_{222}]. \end{array} \tag{7.47}$$

The differential operator for general anisotropic thin plates is, cf. [Lek68],

$$\begin{array}{rrcl} & \mathcal{P}(\partial) &=& [D_{11}\partial_{1111} + 4D_{16}\partial_{1112} + 2(D_{12}+2D_{66})\partial_{1122} \\ & & & +4D_{26}\partial_{1222} + D_{22}\partial_{2222}] \\ \overset{\mathcal{F}}{\leftrightarrow} & \hat{\mathcal{P}}(\hat{x}) &=& [D_{11}\hat{x}_1^4 + 4D_{16}\hat{x}_1^3\hat{x}_2 + 2(D_{12}+2D_{66})\hat{x}_1^2\hat{x}_2^2 \\ & & & +4D_{26}\hat{x}_1\hat{x}_2^3 + D_{22}\hat{x}_2^4]. \end{array} \tag{7.48}$$

Which leads to the Fourier fundamental solution

$$\hat{W} = \frac{1}{D_{11}\hat{x}_1^4 + 4D_{16}\hat{x}_1^3\hat{x}_2 + 2(D_{12}+2D_{66})\hat{x}_1^2\hat{x}_2^2 + 4D_{26}\hat{x}_1\hat{x}_2^3 + D_{22}\hat{x}_2^4}. \tag{7.49}$$

The boundary operators

The relevant boundary operators are in Fourier space

$$\begin{array}{rcl} \hat{\mathcal{A}}^i_\varphi &=& i\nu^i_k\hat{x}_k; \\ \hat{\mathcal{A}}^i_{\mathrm{m}} &=& (\nu^i_1\nu^i_1D_{11} + 2\nu^i_1\nu^i_2D_{16} + \nu^i_2\nu^i_2D_{12})\,\hat{x}_1^2 \\ &+& 2(\nu^i_1\nu^i_1D_{16} + 2\nu^i_1\nu^i_2D_{66} + \nu^i_2\nu^i_2D_{26})\hat{x}_1\hat{x}_2 \\ &+& (\nu^i_1\nu^i_1D_{12} + 2\nu^i_1\nu^i_2D_{26} + \nu^i_2\nu^i_2D_{22})\,\hat{x}_2^2; \\ \hat{\mathcal{A}}^i_{\mathrm{q}} &=& i[\nu^i_1D_{11}(1+\nu^i_2\nu^i_2) - \nu^i_1\nu^i_2\nu^i_2D_{12} + 2\nu^i_2\nu^i_2\nu^i_2D_{16}]\hat{x}_1^3 \\ &+& i[-\nu^i_1\nu^i_1\nu^i_2D_{11} + \nu^i_2(2-\nu^i_2\nu^i_2)D_{12} + 4\nu^i_1D_{16} \\ &-& 2\nu^i_1\nu^i_2\nu^i_2D_{26} + 4\nu^i_2\nu^i_2\nu^i_2D_{66}]\hat{x}_1^2\hat{x}_2 + i[-\nu^i_2\nu^i_2\nu^i_1D_{22} \\ &+& \nu^i_1(2-\nu^i_1\nu^i_1)D_{12} + 4\nu^i_2D_{26} - 2\nu^i_2\nu^i_1\nu^i_1D_{16} + 4\nu^i_1\nu^i_1\nu^i_1D_{66}]\hat{x}_2^2\hat{x}_1 \\ &+& i[\nu^i_2D_{22}(1+\nu^i_1\nu^i_1) - \nu^i_2\nu^i_1\nu^i_1D_{12} + 2\nu^i_1\nu^i_1\nu^i_1D_{26}]\hat{x}_2^3; \\ \mathcal{A}^i_{\mathrm{c}} &=& (\nu_1^{+i}\nu_2^{+i} - \nu_1^{-i}\nu_2^{-i})[(D_{11}-D_{12})\hat{x}_1^2 + (D_{12}-D_{22})\hat{x}_2^2] \\ &+& (\nu_1^{+i}\nu_1^{+i} - \nu_2^{+i}\nu_2^{+i} - \nu_1^{-i}\nu_1^{-i} + \nu_2^{-i}\nu_2^{-i})[D_{16}\hat{x}_1^2 + D_{26}\hat{x}_2^2] \\ &-& [2(\nu_1^{+i}\nu_2^{+i} - \nu_1^{-i}\nu_2^{-i})(D_{16}-D_{26}) + 4\nu_2^{+i}\nu_2^{+i}\nu_2^{-i}\nu_2^{-i}D_{66}]\hat{x}_1\hat{x}_2. \end{array} \tag{7.50}$$

Once more, the Galerkin BIE can be obtained by inserting these operators and the anisotropic fundamental solution into the Galerkin BIE of the isotropic case.

7.1.4 The thin plate on a Winkler foundation

The fundamental solution for the isotropic case

The behavior of an elastically on continuous springs supported plate (Fig.7.5) is described by

$$(D\Delta\Delta + k)w = f \qquad \overset{\mathcal{F}}{\leftrightarrow} \qquad (D(\hat{x}_1^2+\hat{x}_2^2)^2 + k)\hat{w} = \hat{f}. \tag{7.51}$$

It is inverted by the fundamental solution, cf. [Jah98] for the original space,

$$
\begin{aligned}
W &= -\frac{a^2}{2\pi\, D}\,\mathrm{kei}\left(\frac{|x|}{a}\right) = -\frac{a^2}{2\pi\, D}\int_0^\infty \frac{tD}{Dt^4+ka^4}\mathrm{J}_0\left(\frac{t|x|}{a}\right)\,\mathrm{d}t; \\
\overset{\mathcal{F}}{\leftrightarrow}\quad \hat{W} &= \frac{1}{D(\hat{x}_1^2+\hat{x}_2^2)^2+k}, \qquad (7.52)
\end{aligned}
$$

with $a = (D/k)^{1/4}$. kei is the Kelvin function of order zero and J_0 is the Bessel function of order zero. Due to the fact that the differential operator has the

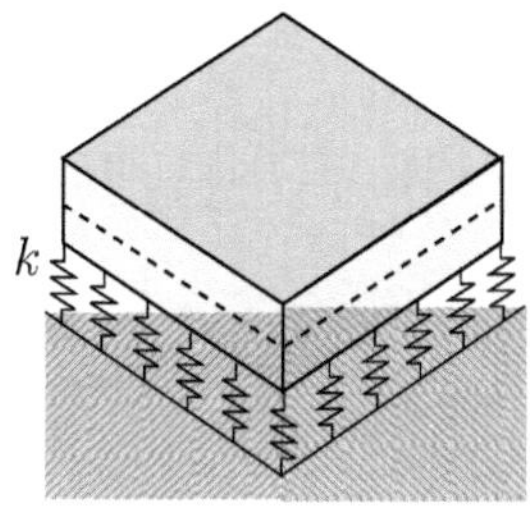

Figure 7.5: *Plate on a Winkler foundation*

same derivations as in the case of the unbedded plate ($k = 0$), the boundary differential operators can be taken from (7.9). The BIE can therefore be adopted from (7.30).

The fundamental solutions for bedded anisotropic plates

The fundamental solutions for the orthotropic and for the general anisotropic case can be taken from the unbedded plate by adding in the denominator the stiffness k. We get for orthotropic plates

$$
\hat{W} = \frac{1}{D_{11}\hat{x}_1^4 + 2D_3\hat{x}_1^2\hat{x}_2^2 + D_{22}\hat{x}_2^4 + k}, \qquad (7.53)
$$

and for general anisotropic plates

$$
\begin{aligned}
\hat{W} &= [D_{11}\hat{x}_1^4 + 4D_{16}\hat{x}_1^3\hat{x}_2 + 2(D_{12}+2D_{66})\hat{x}_1^2\hat{x}_2^2 \qquad (7.54)\\
&\qquad + 4D_{26}\hat{x}_1\hat{x}_2^3 + D_{22}\hat{x}_2^4 + k]^{-1}.
\end{aligned}
$$

7.1.5 Combined bending and stressing of thin plates

The isotropic thin plate

The differential equation for thin plates subjected to compressive in-plane forces N_{kl} is

$$(D\Delta\Delta + N_{kl}\partial_{kl})w = f \qquad \overset{\mathcal{F}}{\leftrightarrow} \qquad (D(\hat{x}_1^2 + \hat{x}_2^2)^2 - N_{kl}\hat{x}_k\hat{x}_l)\hat{w} = \hat{f}. \quad (7.55)$$

The Fourier fundamental solution for arbitrary N_{kl} is

$$\hat{W} = \frac{1}{D(\hat{x}_1^2 + \hat{x}_2^2)^2 - N_{kl}\hat{x}_k\hat{x}_l}. \quad (7.56)$$

For the isotropic case $N_{kl} = N$, [Bes91] gives as fundamental solution in the original space :

$$\text{for tensile } N: \qquad W = \frac{1}{2\pi\zeta^2}\left[\ln|x| + \mathrm{K}_0(\zeta|x|)\right]; \quad (7.57)$$

$$\text{for compressive } N: \qquad W = -\frac{1}{4\pi\zeta^2}\left[2\ln|x| + i\pi\mathrm{H}_0^1(\zeta|x|)\right]. \quad (7.58)$$

The anisotropic thin plate

The Fourier fundamental solutions for orthotropic and general anisotropic plates under membrane forces N_{kl} and on elastic foundations with stiffness k are

$$\text{orthotropic}: \quad \hat{W} = \frac{1}{D_{11}\hat{x}_1^4 + 2D_3\hat{x}_1^2\hat{x}_2^2 + D_{22}\hat{x}_2^4 + k - N_{kl}\hat{x}_k\hat{x}_l}; \quad (7.59)$$

$$\text{anisotropic}: \quad \hat{W} = [D_{11}\hat{x}_1^4 + 4D_{16}\hat{x}_1^3\hat{x}_2 + 2(D_{12} + 2D_{66})\hat{x}_1^2\hat{x}_2^2 \quad (7.60)$$
$$+4D_{26}\hat{x}_1\hat{x}_2^3 + D_{22}\hat{x}_2^4 + k - N_{kl}\hat{x}_k\hat{x}_l]^{-1}.$$

To the author's knowledge, expressions for the original space are not available.

7.2 Refined Plate Theories

7.2.1 The thick plate

To refine Kirchhoffs plate theory Reissner proposed a model where the influence of transverse shearing strains and the effects of the stress σ_{33} perpendicular to the plate are not neglected[4]. The presentation in the original space is essentially based on [AP92]. In the following, we distinguish between Greek indices $\alpha, \beta = 1, 2$ and Latin indices $k, l, m = 1, 2, 3$.

[4]Reissner, E.: On bending of elastic plates. *Quart. Appl. Math.* **5**, (1947).

The differential equation and its fundamental solution

As in the Kirchhoff model, we examine a plate with constant thickness h which is lying in the (x_1, x_2)-plane. $x_3 = 0$ is the mean surface which is subjected to a transverse load $p(x), x = x_1, x_2$ and to volume forces f_k. As generalized displacements u_k we take the rotations $u_1 = \varphi_1, u_2 = \varphi_2$ and the deflection $u_3 = w$ in the direction of x_3. The bending curvatures $\kappa_{\alpha\beta}$ and the transverse rotations of the normals of the plate γ_α are

$$\kappa_{\alpha\beta} = \frac{1}{2}(\partial_\beta\varphi_\alpha + \partial_\alpha\varphi_\beta) \qquad \text{and} \qquad \gamma_\alpha = \varphi_\alpha + \partial_\alpha w. \tag{7.61}$$

These values are related via the constitutive equation to the generalized internal forces

$$\begin{aligned} \bar{m}_{\alpha\beta} &= D(1-\bar{\nu})\left[\kappa_{\alpha\beta} + \frac{\bar{\nu}}{1-\bar{\nu}}\delta_{\alpha\beta}\,\kappa_{\lambda\lambda}\right] + \frac{\bar{\nu}}{1-\bar{\nu}}\delta_{\alpha\beta}\,\lambda^{-2}p; \qquad (7.62)\\ \bar{q}_\alpha &= \lambda^2 D\frac{1-\bar{\nu}}{2}\gamma_\alpha. \end{aligned}$$

$D = (Eh^3)/(12(1-\bar{\nu}^2))$ is the same rigidity as in the Kirchhoff model, cf. (7.2). The characteristic quantity $\lambda^2 = 10/h^2$ related to Reissner's model follows from the assumption of parabolic distribution of the stress σ_{33} along x_3, e.g. [AP92],

$$\sigma_{33} = p(x_1, x_2)\frac{x_3}{2h}\left(3 - \frac{4x_3^2}{h^2}\right). \tag{7.63}$$

It should be noted that the generalized moment $\bar{m}_{\alpha\beta}$ has a term related to the transversal load p. We define $\bar{q}_\alpha = q_\alpha$ and

$$\begin{aligned} \bar{m}_{\alpha\beta} &= m_{\alpha\beta} + \frac{\bar{\nu}}{1-\bar{\nu}}\delta_{\alpha\beta}\,\lambda^{-2}p, \qquad (7.64)\\ \text{thus}: \quad m_{\alpha\beta} &= D(1-\bar{\nu})\left[\kappa_{\alpha\beta} + \frac{\bar{\nu}}{1-\bar{\nu}}\delta_{\alpha\beta}\kappa_{\lambda\lambda}\right]. \end{aligned}$$

The equilibrium equations are

$$\begin{aligned} -q_\alpha + \partial_\beta m_{\alpha\beta} &= -b_\alpha = -f_\alpha - \frac{\bar{\nu}}{1-\bar{\nu}}\lambda^{-2}\partial_\alpha p; \qquad (7.65)\\ \partial_\alpha q_\alpha &= -b_3 = -f_3 - p. \end{aligned}$$

We get finally the partial differential equation

$$\mathcal{P}_{kl}(\partial)u_l(x) = -b_k(x) \qquad \overset{\mathcal{F}}{\leftrightarrow} \qquad \hat{\mathcal{P}}_{kl}(\hat{x})\hat{u}_l(\hat{x}) = -\hat{b}_k(\hat{x}), \tag{7.66}$$

and the differential operator

$$\mathcal{P}(\partial) = \frac{D(1-\bar{\nu})}{2}\begin{bmatrix} (\Delta-\lambda^2)+c_1\partial_{11} & c_1\partial_{12} & -\lambda^2\partial_1 \\ c_1\partial_{21} & (\Delta-\lambda^2)+c_1\partial_{22} & -\lambda^2\partial_2 \\ \lambda^2\partial_1 & \lambda^2\partial_2 & \lambda^2\Delta \end{bmatrix} \tag{7.67}$$

$$\hat{\mathcal{P}}(\hat{x}) = \frac{D(\bar{\nu}-1)}{2}\begin{bmatrix} (|\hat{x}|^2+\lambda^2)+c_1\hat{x}_1^2 & c_1\hat{x}_1\hat{x}_2 & i\lambda^2\hat{x}_1 \\ c_1\hat{x}_2\hat{x}_1 & (|\hat{x}|^2+\lambda^2)+c_1\hat{x}_2^2 & i\lambda^2\hat{x}_2 \\ -i\lambda^2\hat{x}_1 & -i\lambda^2\hat{x}_2 & \lambda^2|\hat{x}|^2 \end{bmatrix}$$

with $c_1 = (1+\bar{\nu})/(1-\bar{\nu})$ and $\Delta = \partial_{11}+\partial_{22}$, $|\hat{x}|^2 = \hat{x}_1^2+\hat{x}_2^2$. The right-hand side is with $c_2 = \bar{\nu}/(1-\bar{\nu})$

$$(b)^T = (\ f_1 + c_2\lambda^{-2}\partial_1 p,\quad f_2 + c_2\lambda^{-2}\partial_2 p,\quad f_3 - p\ ). \tag{7.68}$$

The inverse leads to the Fourier fundamental solution :

$$\hat{U} = \frac{2}{D(\bar{\nu}-1)}\begin{bmatrix} (|\hat{x}|^2+\lambda^2)+c_1\hat{x}_1^2 & c_1\hat{x}_1\hat{x}_2 & i\lambda^2\hat{x}_1 \\ c_1\hat{x}_2\hat{x}_1 & (|\hat{x}|^2+\lambda^2)+c_1\hat{x}_2^2 & i\lambda^2\hat{x}_2 \\ -i\lambda^2\hat{x}_1 & -i\lambda^2\hat{x}_2 & \lambda^2|\hat{x}|^2 \end{bmatrix}^{-1} \tag{7.69}$$

The version in the original space is given for example in [Ant88] and reproduced here

$$\begin{aligned} U_{\alpha\beta} &= \frac{1}{8\pi D}\left[\left(\frac{8B(z)}{1-\bar{\nu}} - \ln z^2 + 1\right)\delta_{\alpha\beta} - \left(\frac{8A(z)}{1-\bar{\nu}}+2\right)\partial_\alpha|x|\partial_\beta|x|\right]; \\ U_{\alpha 3} &= -U_{3\alpha} = \frac{1}{8\pi D}|x|\partial_\alpha|x|\left(\ln z^2 - 1\right); \\ U_{33} &= \frac{1}{16\pi D\lambda^2}\left[\left(z^2 - \frac{8}{1-\bar{\nu}}\right)\ln z^2 - 2z^2\right), \end{aligned} \tag{7.70}$$

with $|x| = \sqrt{x_1^2+x_2^2}$, $z = \lambda|x|$ and

$$A(z) = \mathrm{K}_0(z) + \frac{2}{z}\left(\mathrm{K}_1(z) - \frac{1}{z}\right) \quad \text{and} \quad B(z) = \mathrm{K}_0(z) + \frac{1}{z}\left(\mathrm{K}_1(z) - \frac{1}{z}\right).$$

$\mathrm{K}_0, \mathrm{K}_1$ are the modified Bessel functions.

The Galerkin boundary integral equations

The $\hat{U}$ defined in (7.69) are the fundamental quantities in the field. The corresponding boundary terms (normal slope, tangential slope, and displacement) $\hat{V}_1^i = (\hat{\Phi}_\nu, \hat{\Phi}_\tau, \hat{W})$ at the i-th part of the boundary with the normal

$\nu^i = (\nu_1^i, \nu_2^i)$ are obtained by

$$\hat{V}_1^i = \hat{\mathcal{N}}^i \hat{U}, \tag{7.71}$$
$$\text{with :} \quad \hat{\mathcal{N}}^i = \begin{bmatrix} \nu_1^i & \nu_2^i & 0 \\ -\nu_2^i & \nu_1^i & 0 \\ 0 & 0 & 1 \end{bmatrix}.$$

The transform of the boundary differential operator $\mathcal{A}_\nu^i$ to obtain the missing fundamental quantities $\hat{V}_2^i = (\hat{M}_{\nu\nu}^i, \hat{M}_{\nu\tau}^i, \hat{Q}_\nu^i)$ (the bending moment, the twisting moment, and the real shearing force) is, cf. [AHFEZ99],

$$\hat{V}_2^i = \hat{\mathcal{A}}_\nu^i \hat{V}_1^i, \tag{7.72}$$
$$\hat{\mathcal{A}}_\nu^i = \begin{bmatrix} i\nu_k^i \hat{x}_k & i\bar{\nu}\tau_k^i \hat{x}_k & 0 \\ \frac{1-\bar{\nu}}{2} i\bar{\nu}\tau_k^i \hat{x}_k & \frac{1-\bar{\nu}}{2} i\bar{\nu}\nu_k^i \hat{x}_k & 0 \\ \frac{1-\bar{\nu}}{2}\lambda^2 & 0 & \frac{1-\bar{\nu}}{2}\lambda^2 i\nu_k^i \hat{x}_k \end{bmatrix}$$

with $\tau^i = (\tau_1^i, \tau_2^i)$ as the i-th tangential vector.

Because the differential operator is of order two and not of order four as in the Kirchhoff model, linear shape functions $\phi_{\nu,\tau,\mathrm{w}}^i$ are sufficient for the boundary "displacements" $\varphi_\nu, \varphi_\tau, w$, whereas constant functions $\phi_{\mathrm{m,t,q}}^i$ can be taken for the derivatives $m_{\nu\nu}, m_{\nu\tau}, q_\nu$. The discretizations are :

$$\varphi_\nu \approx \sum_i^{N_\nu} \mathfrak{n}^i \phi_\nu^i(x), \quad \varphi_\tau \approx \sum_i^{N_\tau} \mathfrak{s}^i \phi_\tau^i(x), \quad w \approx \sum_i^{N_\mathrm{w}} \mathfrak{w}^i \phi_\mathrm{w}^i(x), \tag{7.73}$$
$$m_{\nu\nu} \approx \sum_i^{N_\mathrm{m}} \mathfrak{m}^i \phi_\mathrm{m}^i(x), \quad m_{\nu\tau} \approx \sum_i^{N_\mathrm{t}} \mathfrak{t}^i \phi_\mathrm{t}^i(x), \quad q_\nu \approx \sum_i^{N_\mathrm{q}} \mathfrak{q}^i \phi_\mathrm{q}^i(x).$$

The boundary integral equation is

$$\left\langle \hat{\phi}^j(-\hat{x}), \hat{\mathcal{B}}\hat{u}_\chi \right\rangle = \left\langle \hat{\phi}^j(-\hat{x}), \hat{f}_\chi \hat{\mathcal{B}}\hat{V} \right\rangle + \sum_i \mathfrak{u}^i \left\langle \hat{\phi}^j(-\hat{x}), \hat{\phi}^i \hat{\mathcal{A}}\hat{V} \right\rangle, \tag{7.74}$$

with

$$\hat{\mathcal{A}} = \begin{bmatrix} \mathcal{I} & -\hat{\mathcal{A}}_\nu^i \\ -\hat{\mathcal{A}}_\nu^j & \hat{\mathcal{A}}_\nu^j \hat{\mathcal{A}}_\nu^i \end{bmatrix}; \quad \hat{\mathcal{B}} = \begin{pmatrix} \mathcal{I} \\ \hat{\mathcal{A}}_\nu^j \end{pmatrix},$$

7.2.2 Thick plates on Winkler foundations

The fundamental solution

The substitution $p \to p - kw$ in (7.65) leads to the equilibrium equation for a thick plate on a Winkler foundation (k is the modulus of subgrade reaction). The transformed differential operator is

$$\hat{\mathcal{P}}(\hat{x}) = \frac{D(\bar{\nu}-1)}{2}\begin{bmatrix} (|\hat{x}|^2+\lambda^2)+c_1\hat{x}_1^2 & c_1\hat{x}_1\hat{x}_2 & i\lambda^2\hat{x}_1 - ic_2\hat{x}_1 \\ c_1\hat{x}_2\hat{x}_1 & (|\hat{x}|^2+\lambda^2)+c_1\hat{x}_2^2 & i\lambda^2\hat{x}_2 - ic_2\hat{x}_2 \\ -i\lambda^2\hat{x}_1 & -i\lambda^2\hat{x}_2 & \lambda^2|\hat{x}|^2 - c_3 \end{bmatrix}$$

with $c_2 = 2k\bar{\nu}/(D(1-\bar{\nu})^2\lambda^2)$ and $c_3 = 2k/(D(1-\bar{\nu}))$. The fundamental solution is the inverse

$$\hat{U} = \hat{\mathcal{P}}^{-1}(\hat{x}). \tag{7.75}$$

Chapter 8

Waves

Lord Rayleigh

8.1 Transient Problems

8.1.1 Fourier BIE for transient problems

In the context of transient dynamic problems the symbol $\hat{(\cdot)}$ is used to denote a spatial and temporal Fourier transformed quantity, i.e. we define in analogy to (4.1) :

$$\hat{u}(\hat{x}, \omega) = \int_{\mathbb{R}} \int_{\mathbb{R}^n} u(x,t)\; \mathrm{e}^{-i<x,\hat{x}>} \mathrm{e}^{-i\omega t}\, \mathrm{d}x\, \mathrm{d}t. \tag{8.1}$$

$t, x_k, k = 1, 2, 3$ are the time and the spatial coordinates. The corresponding differentiations are denoted by ∂_t, ∂_k. The quantities $\omega, \hat{x}_k$ are the circular frequency and the wave numbers. For the dynamic Fourier BEM the scalar product and the convolution have to be redefined, cf. Eqs. (3.25) and (3.26),

$$\text{scalar product:} \qquad \langle a, b \rangle = \int_{\mathbb{R}} \int_{\mathbb{R}^n} a(x,t)\; b(x,t)\, \mathrm{d}x\, \mathrm{d}t; \tag{8.2}$$

$$\text{convolution:} \qquad a * b = \int_{\mathbb{R}} \int_{\mathbb{R}^n} a(y,\tau)\; b(x-y, t-\tau)\, \mathrm{d}y\, \mathrm{d}\tau. \tag{8.3}$$

The convolution theorem combined with Parseval's equality lead to the fundamental equality for the dynamic Fourier BEM

$$\langle a(x,t), b(x,t) * c(x,t) \rangle = \frac{1}{(2\pi)^{n+1}} \left\langle \hat{a}(-\hat{x}, -\omega), \hat{b}(\hat{x}, \omega)\, \hat{c}(\hat{x}, \omega) \right\rangle. \tag{8.4}$$

Literature for the traditional BEM can be found in [Ant88, Bes87, Bes97, Bon99].

8.1.2 The elastic bar

The differential equation and its fundamental solution

To illustrate the general principle of Fourier BEM for transient problems the example of an elastic bar is taken as prototype. It is a two-dimensional problem in $\mathbb{R}^1 \times \mathbb{R}^+$.

The differential equation and its Fourier transform describing the dynamic behavior of the bar are

$$\begin{aligned} (-EA\,\partial_1^2 + \rho\,\partial_t^2)\,u(x_1,t) &= f(x_1,t) \\ \overset{\mathcal{F}}{\leftrightarrow} \quad (EA\,\hat{x}_1^2 - \rho\,\omega^2)\,\hat{u}(\hat{x}_1,\omega) &= \hat{f}(\hat{x}_1,\omega), \end{aligned} \tag{8.5}$$

with EA, ρ as the longitudinal rigidity and the mass density, respectively. u is the longitudinal displacement and f is the volume force in x_1 direction. The traction q at the boundary with normal ν^i is[1]

$$q^i = \mathcal{A}_t^i u = EA\,\partial_1 u \qquad \overset{\mathcal{F}}{\leftrightarrow} \qquad \hat{q}^i = \hat{\mathcal{A}}_t^i \hat{u} = iEA\,\hat{x}_1\hat{u}. \tag{8.6}$$

The boundary consists only of the two end points of the bar. The fundamental solution and its derivative are obtained by setting $f = \delta(x_1)\delta(t) \overset{\mathcal{F}}{\leftrightarrow} \hat{f} = 1$:

$$\begin{aligned} U &= \frac{1}{4}\mathrm{sgn}\,(t)\,[\mathrm{H}(-x_1 - c_\mathrm{p}t) - \mathrm{H}(-x_1 + c_\mathrm{p}t)] \\ \overset{\mathcal{F}}{\leftrightarrow} \quad \hat{U} &= \frac{1}{EA\hat{x}_1^2 - \rho\,\omega^2}, \end{aligned} \tag{8.7}$$

with $c_\mathrm{p} = \sqrt{EA/\rho}$ as wave velocity.

Causal fundamental solution

Obviously, the fundamental solution is not causal, a wave is traveling in direction of negative time. As proposed in [Dud97b], pp 58, a simple residual calculus, based on the theorem of Paley-Wiener, renders for the causal fundamental solution

$$\begin{aligned} U &= \frac{1}{2}\mathrm{H}(t)\,[\mathrm{H}(-x_1 - c_\mathrm{p}t) - \mathrm{H}(-x_1 + c_\mathrm{p}t)] \\ \overset{\mathcal{F}}{\leftrightarrow} \quad \hat{U} &= \frac{1}{EA\hat{x}_1^2 - \rho\,\omega^2} + \frac{i\rho\pi}{2c_\mathrm{p}|\hat{x}_1|}\left[\delta(\omega - c_\mathrm{p}\hat{x}_1) - \delta(\omega + c_\mathrm{p}\hat{x}_1)\right]. \end{aligned} \tag{8.8}$$

[1]Compared to the static examples the notation for the traction t is changed here to q to avoid confusion with t as time.

Dynamic Galerkin BIE for the bar

The boundary quantities u, q are approximated by spatial $\phi^{i_1}_{\mathrm{q,u}}(x_1)$ and temporal trial functions $\varphi^{i_2}_{\mathrm{q,u}}(t)$

$$u(x_1,t) \approx \sum_{i_1,i_2} \mathrm{u}^{i_1 i_2}\, \phi^{i_1}_{\mathrm{u}}(x_1)\varphi^{i_2}_{\mathrm{u}}(t) \;\overset{\mathcal{F}}{\leftrightarrow}\; \hat{u}(\hat{x}_1,\omega) \approx \sum_{i_1,i_2} \mathrm{u}^{i_1 i_2}\, \hat{\phi}^{i_1}_{\mathrm{u}}(\hat{x}_1)\hat{\varphi}^{i_2}_{\mathrm{u}}(\omega); \quad (8.9)$$

$$q(x_1,t) \approx \sum_{i_1,i_2} \mathrm{q}^{i_1 i_2}\, \phi^{i_1}_{\mathrm{q}}(x_1)\,\varphi^{i_2}_{\mathrm{q}}(t) \;\overset{\mathcal{F}}{\leftrightarrow}\; \hat{q}(\hat{x}_1,\omega) \approx \sum_{i_1,i_2} \mathrm{q}^{i_1 i_2}\, \hat{\phi}^{i_1}_{\mathrm{q}}(\hat{x}_1)\,\hat{\varphi}^{i_2}_{\mathrm{q}}(\omega).$$

Temporal and spatial test functions $\phi^{j_1}_{\mathrm{q}}(x), \varphi^{j_2}_{\mathrm{q}}(t)$ lead to the displacement Galerkin BIE, cf. [Bar99] for mathematical proofs,

$$\begin{aligned} \langle \phi^{j_1}_{\mathrm{q}}\varphi^{j_2}_{\mathrm{q}}, u_\chi \rangle &= \langle \phi^{j_1}_{\mathrm{q}}\varphi^{j_2}_{\mathrm{q}}, f_\chi * U \rangle + \sum_{i_1,i_2} \mathrm{q}^{i_1 i_2} \langle \phi^{j_1}_{\mathrm{q}}\varphi^{j_2}_{\mathrm{q}}, \phi^{i_1}_{\mathrm{q}}\,\varphi^{i_2}_{\mathrm{q}} * U \rangle \qquad (8.10) \\ &- \sum_{i_1,i_2} \mathrm{u}^{i_1 i_2} \langle \phi^{j_1}_{\mathrm{q}}\,\varphi^{j_2}_{\mathrm{q}}, \phi^{i_1}_{\mathrm{u}}\,\varphi^{i_2}_{\mathrm{u}} * \mathcal{A}^i_{\mathrm{t}} U \rangle . \end{aligned}$$

Once more, the free term κ is hidden in the distributional notation. All quantities are formally extended from their bounded supports to $\mathbb{R}^1 \times \mathbb{R}^1$ as shown in Chapter 3. The equivalent BIE is in Fourier space

$$\begin{aligned} \left\langle \hat{\phi}^{j_1}_{\mathrm{q}}(-\hat{x}_1)\hat{\varphi}^{j_2}_{\mathrm{q}}(-\omega), \hat{u}_\chi \right\rangle &= \left\langle \hat{\phi}^{j_1}_{\mathrm{q}}(-\hat{x}_1)\hat{\varphi}^{j_2}_{\mathrm{q}}(-\omega), \hat{f}_\chi \hat{U} \right\rangle \qquad (8.11) \\ &+ \sum_{i_1,i_2} \mathrm{q}^{i_1 i_2} \left\langle \hat{\phi}^{j_1}_{\mathrm{q}}(-\hat{x}_1)\hat{\varphi}^{j_2}_{\mathrm{q}}(-\omega), \hat{\phi}^{i_1}_{\mathrm{q}}\hat{\varphi}^{i_2}_{\mathrm{q}}\hat{U} \right\rangle \\ &- \sum_{i_1,i_2} \mathrm{u}^{i_1 i_2} \left\langle \hat{\phi}^{j_1}_{\mathrm{q}}(-\hat{x}_1)\hat{\varphi}^{j_2}_{\mathrm{q}}(-\omega), \hat{\phi}^{i_1}_{\mathrm{u}}\hat{\varphi}^{i_2}_{\mathrm{u}}\hat{\mathcal{A}}^i_{\mathrm{t}}\hat{U} \right\rangle , \end{aligned}$$

where the factor $(2\pi)^{-3}$ was cancelled. Its analogue, the traction BIE is

$$\begin{aligned} -\left\langle \hat{\phi}^{j_1}_{\mathrm{u}}(-\hat{x}_1)\hat{\varphi}^{j_2}_{\mathrm{u}}(-\omega), \hat{\mathcal{A}}^j_{\mathrm{t}}\hat{u}_\chi \right\rangle &= -\left\langle \hat{\phi}^{j_1}_{\mathrm{u}}(-\hat{x}_1)\hat{\varphi}^{j_2}_{\mathrm{u}}(-\omega), \hat{f}_\chi \hat{\mathcal{A}}^j_{\mathrm{t}}\hat{U} \right\rangle \qquad (8.12) \\ &- \sum_{i_1,i_2} \mathrm{q}^{i_1 i_2} \left\langle \hat{\phi}^{j_1}_{\mathrm{u}}(-\hat{x}_1)\hat{\varphi}^{j_2}_{\mathrm{u}}(-\omega), \hat{\phi}^{i_1}_{\mathrm{q}}\hat{\varphi}^{i_2}_{\mathrm{q}}\hat{\mathcal{A}}^j_{\mathrm{t}}\hat{U} \right\rangle \\ &+ \sum_{i_1,i_2} \mathrm{u}^{i_1 i_2} \left\langle \hat{\phi}^{j_1}_{\mathrm{u}}(-\hat{x}_1)\hat{\varphi}^{j_2}_{\mathrm{u}}(-\omega), \hat{\phi}^{i_1}_{\mathrm{u}}\hat{\varphi}^{i_2}_{\mathrm{u}}\hat{\mathcal{A}}^j_{\mathrm{t}}\hat{\mathcal{A}}^i_{\mathrm{t}}\hat{U} \right\rangle , \end{aligned}$$

For this example, the formal presentation is exaggerated but facilitates the transfer to problems which are more complex.

The elastic bar under transient loading
As an example, we examine a transient volume force

$$f = \delta(x_1 - 3/2)[\mathrm{H}(t) - \mathrm{H}(t-1/2)] \quad \overset{\mathcal{F}}{\leftrightarrow} \quad \hat{f} = \frac{1 - \mathrm{e}^{-i\omega/2}}{i\omega}\,\mathrm{e}^{-3i\hat{x}_1/2}, \quad (8.13)$$

which is placed at the quarter point of the bar and lasts from $t = 0$ to $t = 1/2$ [s], cf. Fig.8.1. A wave velocity of $c_\mathrm{p} = \sqrt{1/4}$ [m/s] and a bar length of $L = 2$ [m] are chosen. As spatial trial and test functions we have

$$\begin{array}{llcll} \text{for } u(L): & \phi^1_\mathrm{u}(x_1) = \delta(x_1 - L) & \overset{\mathcal{F}}{\leftrightarrow} & \hat{\phi}^1_\mathrm{u}(\hat{x}_1) = \mathrm{e}^{-iL\hat{x}_1} \\ \text{for } q(0): & \phi^1_\mathrm{q}(x_1) = \delta(x_1) & \overset{\mathcal{F}}{\leftrightarrow} & \hat{\phi}^1_\mathrm{q}(\hat{x}_1) = 1. \end{array}$$

The linear temporal trial and test functions φ^{j_2} for the j_2-th time step are constructed by translations $t \to t - j_2 t_\triangle$ in the original domain and by modulation with $\mathrm{e}^{-ij_2 t_\triangle \omega}$ in the transformed domain of the reference element

$$\begin{aligned} \varphi^0 &= \frac{t + t_\triangle}{t_\triangle}[\mathrm{H}(t + t_\triangle) - \mathrm{H}(t)] + \frac{t - t_\triangle}{t_\triangle}[\mathrm{H}(t) - \mathrm{H}(t - t_\triangle)] \\ \overset{\mathcal{F}}{\leftrightarrow} \quad \hat{\varphi}_0 &= \frac{1 - \cos(\omega t_\triangle)}{2 t_\triangle \omega^2}. \end{aligned}$$

A time step size of $t_\triangle = 1/10$ [1/s] is chosen (half the support of one time trial function) and about hundred time steps were calculated. With homogeneous initial conditions and with boundary conditions $u(0,t) = 0, q(L,t) = 0$ we get the displacements and tractions at $u(L,t), q(0,t)$, cf. Fig.8.2, which shows clearly the reflections of the propagating wave.

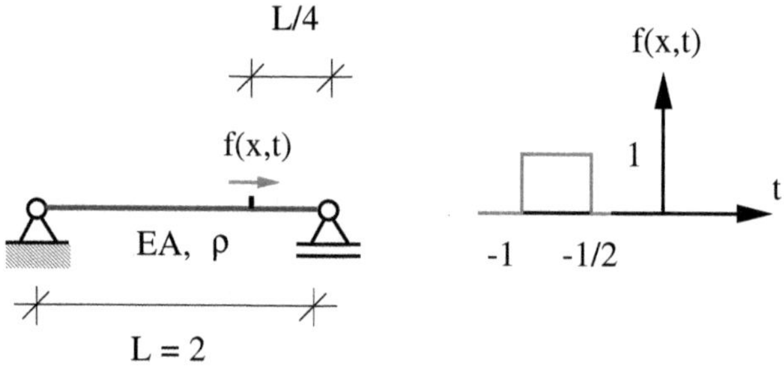

Figure 8.1: *Geometry and loading of the elastic bar*

Evidently, this simple example can be solved by the traditional BEM approach. It is chosen here to clarify the procedure for cases where the standard BEM can not be established, i.e. where fundamental solutions are unknown.

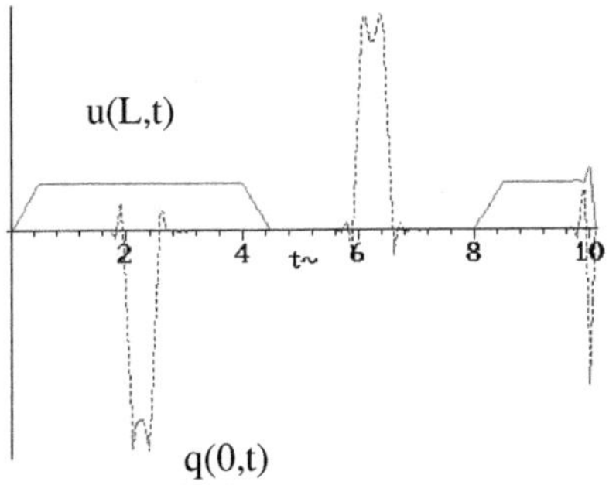

Figure 8.2: *Boundary displacement $u(L,t)$ and boundary traction $q(0,t)$ for the elastic bar*

8.1.3 Scalar waves in isotropic media

The fundamental solution and the Galerkin BIE

Simple wave problems (acoustic waves, water waves, etc.) can be described by the scalar wave equation in $\mathbb{R}^2 \times \mathbb{R}^+$ or $\mathbb{R}^3 \times \mathbb{R}^+$

$$\left(\Delta - \frac{1}{c_\mathrm{p}^2}\partial_{tt}\right) u(x,t) = -f(x,t) \overset{\mathcal{F}}{\leftrightarrow} \left(-|\hat{x}|^2 + \frac{\omega^2}{c_\mathrm{p}^2}\right) \hat{u}(\hat{x},\omega) = -\hat{f}(\hat{x},\omega). \quad (8.14)$$

u is the velocity potential, c_p is the known acoustic velocity characteristic for the medium, and $\mathcal{A}_\mathrm{t}^i = \nu\nabla$ is the boundary differential operator. The fundamental solutions are for $\mathbb{R}^{2,3}$ (cf. [Bon95a] for the original space)

$$\text{for } \mathbb{R}^2: \quad U = \frac{\operatorname{sgn}(c_\mathrm{p}t - |x|)}{4\pi c_\mathrm{p}\sqrt{c_\mathrm{p}^2 t^2 - |x|^2}} \quad \overset{\mathcal{F}}{\leftrightarrow} \quad \hat{U} = \frac{c_\mathrm{p}^2}{\omega^2 - c_\mathrm{p}^2|\hat{x}|^2}; \quad (8.15)$$

$$\text{for } \mathbb{R}^3: \quad U = \frac{\delta(t - |x|/c_\mathrm{p})}{4\pi|x|} \quad \overset{\mathcal{F}}{\leftrightarrow} \quad \hat{U} = \frac{c_\mathrm{p}^2}{\omega^2 - c_\mathrm{p}^2|\hat{x}|^2}. \quad (8.16)$$

The causality in Fourier space is achieved as discussed in Section 8.1.2 :

$$\begin{aligned}
\text{for the } \mathbb{R}^2: \quad U &= \frac{\mathrm{H}(c_\mathrm{p}t - |x|)}{2\pi c_\mathrm{p}\sqrt{c_\mathrm{p}^2 t^2 - |x|^2}} \qquad (8.17)\\
\overset{\mathcal{F}}{\leftrightarrow} \quad \hat{U} &= \frac{c_\mathrm{p}^2}{\omega^2 - c_\mathrm{p}^2|\hat{x}|^2} + \frac{i\pi}{2c_\mathrm{p}|\hat{x}|}\left[\delta(\omega - c_\mathrm{p}|\hat{x}|) - \delta(\omega + c_\mathrm{p}|\hat{x}|)\right];\\
\text{for the } \mathbb{R}^3: \quad U &= \frac{\delta(t - |x|/c_\mathrm{p})}{4\pi|x|}\mathrm{H}(t)\\
\overset{\mathcal{F}}{\leftrightarrow} \quad \hat{U} &= \frac{c_\mathrm{p}^2}{\omega^2 - c_\mathrm{p}^2|\hat{x}|^2} + \frac{i\pi}{2c_\mathrm{p}|\hat{x}|}\left[\delta(\omega - c_\mathrm{p}|\hat{x}|) - \delta(\omega + c_\mathrm{p}|\hat{x}|)\right].
\end{aligned}$$

In Fourier space, the Galerkin BIE equivalent to (8.14) is, compare (8.11) and (8.12),

$$\begin{aligned}
\left\langle \hat{\phi}_\mathrm{q}^{j_1}(-\hat{x})\hat{\varphi}_\mathrm{q}^{j_2}(-\omega), \hat{u}_\chi \right\rangle &= \left\langle \hat{\phi}_\mathrm{q}^{j_1}(-\hat{x})\hat{\varphi}_\mathrm{q}^{j_2}(-\omega), \hat{f}_\chi \hat{U} \right\rangle \qquad (8.18)\\
&+ \sum_{i_1,i_2} \mathsf{q}^{i_1 i_2} \left\langle \hat{\phi}_\mathrm{q}^{j_1}(-\hat{x})\hat{\varphi}_\mathrm{q}^{j_2}(-\omega), \hat{\phi}_\mathrm{q}^{i_1}\hat{\varphi}_\mathrm{q}^{i_2}\hat{U} \right\rangle\\
&- \sum_{i_1,i_2} \mathsf{u}^{i_1 i_2} \left\langle \hat{\phi}_\mathrm{q}^{j_1}(-\hat{x})\hat{\varphi}_\mathrm{q}^{j_2}(-\omega), \hat{\phi}_\mathrm{q}^{i_1}\hat{\varphi}_\mathrm{q}^{i_2}\hat{\mathcal{A}}_\mathrm{t}^{i}\hat{U} \right\rangle,
\end{aligned}$$

and for the traction BIE we get

$$\begin{aligned}
-\left\langle \hat{\phi}_\mathrm{u}^{j_1}(-\hat{x})\hat{\varphi}_\mathrm{u}^{j_2}(-\omega), \hat{\mathcal{A}}_\mathrm{t}^{j}\hat{u}_\chi \right\rangle &= -\left\langle \hat{\phi}_\mathrm{u}^{j_1}(-\hat{x})\hat{\varphi}_\mathrm{u}^{j_2}(-\omega), \hat{f}_\chi \hat{\mathcal{A}}_\mathrm{t}^{j}\hat{U} \right\rangle \qquad (8.19)\\
&- \sum_{i_1,i_2} \mathsf{q}^{i_1 i_2} \left\langle \hat{\phi}_\mathrm{u}^{j_1}(-\hat{x})\hat{\varphi}_\mathrm{u}^{j_2}(-\omega), \hat{\phi}_\mathrm{q}^{i_1}\hat{\varphi}_\mathrm{q}^{i_2}\hat{\mathcal{A}}_\mathrm{t}^{j}\hat{U} \right\rangle\\
&+ \sum_{i_1,i_2} \mathsf{u}^{i_1 i_2} \left\langle \hat{\phi}_\mathrm{u}^{j_1}(-\hat{x})\hat{\varphi}_\mathrm{u}^{j_2}(-\omega), \hat{\phi}_\mathrm{u}^{i_1}\hat{\varphi}_\mathrm{u}^{i_2}\hat{\mathcal{A}}_\mathrm{t}^{j}\hat{\mathcal{A}}_\mathrm{t}^{i}\hat{U} \right\rangle.
\end{aligned}$$

8.1.4 Waves in isotropic elastic media

The fundamental solution and the Galerkin BIE

The dynamic behaviour of the isotropic, homogeneous elastic continuum is described by Lamé's equation

$$\begin{aligned}
\mu\,\partial_{ll}u_k + (\lambda + \mu)\,\partial_{kl}u_l - \rho\,\partial_{tt}u_k &= -f_k, \qquad k, l = 1, 2, 3;\\
\overset{\mathcal{F}}{\leftrightarrow} \quad -\mu\,\hat{x}_l\hat{x}_l\hat{u}_k - (\lambda + \mu)\,\hat{x}_k\hat{x}_l\hat{u}_l + \rho\,\omega^2\hat{u}_k &= -\hat{f}_k. \qquad (8.20)
\end{aligned}$$

which is an extension of Navier's equation (6.1) by the inertial term $\rho\partial_{tt}u_k$ (ρ is the mass density). In analogy to (6.4) the dynamic differential operator is obtained as

$$\mathcal{P}(\partial) = \begin{bmatrix} c_s^2\Delta + c_1^2\partial_{11} - \partial_{tt} & c_1^2\partial_{12} & c_1^2\partial_{13} \\ c_1^2\partial_{21} & c_s^2\Delta + c_1^2\partial_{22} - \partial_{tt} & c_1^2\partial_{23} \\ c_1^2\partial_{31} & c_1^2\partial_{32} & c_s^2\Delta + c_1^2\partial_{33} - \partial_{tt} \end{bmatrix} \tag{8.21}$$

Its transform is

$$\hat{\mathcal{P}}(\hat{x}) = -\begin{bmatrix} c_s^2|\hat{x}|^2 + c_1^2\hat{x}_1^2 - \omega^2 & c_1^2\hat{x}_1\hat{x}_2 & c_1^2\hat{x}_1\hat{x}_3 \\ c_1^2\hat{x}_2\hat{x}_1 & c_s^2|\hat{x}|^2 + c_1^2\hat{x}_2^2 - \omega^2 & c_1^2\hat{x}_2\hat{x}_3 \\ c_1^2\hat{x}_3\hat{x}_1 & c_1^2\hat{x}_3\hat{x}_2 & c_s^2|\hat{x}|^2 + c_1^2\hat{x}_3^2 - \omega^2 \end{bmatrix} \tag{8.22}$$

and hence its Fourier fundamental solution

$$\hat{U} = \begin{bmatrix} c_s^2|\hat{x}|^2 + c_1^2\hat{x}_1^2 - \omega^2 & c_1^2\hat{x}_1\hat{x}_2 & c_1^2\hat{x}_1\hat{x}_3 \\ c_1^2\hat{x}_2\hat{x}_1 & c_s^2|\hat{x}|^2 + c_1^2\hat{x}_2^2 - \omega^2 & c_1^2\hat{x}_2\hat{x}_3 \\ c_1^2\hat{x}_3\hat{x}_1 & c_1^2\hat{x}_3\hat{x}_2 & c_s^2|\hat{x}|^2 + c_1^2\hat{x}_3^2 - \omega^2 \end{bmatrix}^{-1} \tag{8.23}$$

For two dimensional problems we have to take the upper left part of this matrix. The traction boundary operator $\mathcal{A}_{\mathrm{q}}^i$ is identical to that of the static case $\mathcal{A}_{\mathrm{t}}^i$, see (6.15). In for example [ES75] the three-dimensional fundamental solution was derived for the original space

$$\begin{aligned} U_{kl} &= \frac{t}{4\pi\rho|x|^2}\left\{\frac{1}{|x|}(3\partial_k|x|\partial_l|x| - \delta_{kl})\left[\frac{1}{c_{\mathrm{p}}}\mathrm{H}(c_{\mathrm{p}}t - |x|) - \frac{1}{c_{\mathrm{s}}}\mathrm{H}(c_{\mathrm{s}}t - |x|)\right]\right. \\ &+ \left.\partial_k|x|\partial_l|x|\left[\frac{1}{c_{\mathrm{p}}^2}\delta(c_{\mathrm{p}}t - |x|) - \frac{1}{c_{\mathrm{s}}^2}\delta(c_{\mathrm{s}}t - |x|)\right] + \frac{\delta_{kl}}{c_{\mathrm{s}}^2}\delta(c_{\mathrm{s}}t - |x|)\right\} \end{aligned} \tag{8.24}$$

And for two dimensions we have, e.g. [Kyt96],

$$\begin{aligned} U_{kl} &= \frac{1}{4\pi\rho|x|^2}\left\{\frac{1}{c_{\mathrm{p}}}\mathrm{H}(c_{\mathrm{p}}t - |x|)\left[\frac{2c_{\mathrm{p}}^2t^2 - |x|^2}{R_{\mathrm{p}}}\partial_k|x|\partial_l|x| - R_{\mathrm{p}}\delta_{kl}\right]\right. \\ &- \left.\frac{1}{c_{\mathrm{s}}}\mathrm{H}(c_{\mathrm{s}}t - |x|)\left[\frac{2c_{\mathrm{s}}^2t^2 - |x|^2}{R_{\mathrm{s}}}\partial_k|x|\partial_l|x| - \left(R_{\mathrm{s}} + \frac{|x|^2}{R_{\mathrm{s}}^2}\right)\delta_{kl}\right]\right\}, \end{aligned} \tag{8.25}$$

with $R_{\mathrm{p,s}} = \sqrt{c_{\mathrm{p,s}}^2t^2 - |x|^2}$.

The Galerkin BIE

The vectorial Galerkin BIE for general transient dynamic problems is (cf. [MDC89] for the original space) in Fourier space

$$\begin{aligned}\left\langle \hat{\phi}_{\mathrm{q}}^{j_1}(-\hat{x})\hat{\varphi}_{\mathrm{q}}^{j_2}(-\omega), \hat{u}_\chi \right\rangle = &\left\langle \hat{\phi}_{\mathrm{q}}^{j_1}(-\hat{x})\hat{\varphi}_{\mathrm{q}}^{j_2}(-\omega), \hat{f}_\chi \hat{U} \right\rangle \\ &+ \sum_{i_1,i_2} \mathsf{q}^{i_1 i_2} \left\langle \hat{\phi}_{\mathrm{q}}^{j_1}(-\hat{x})\hat{\varphi}_{\mathrm{q}}^{\mathrm{j}_2}(-\omega), \hat{\phi}_{\mathrm{q}}^{i_1}\hat{\varphi}_{\mathrm{q}}^{i_2}\hat{U} \right\rangle \\ &- \sum_{i_1,i_2} \mathsf{u}^{i_1 i_2} \left\langle \hat{\phi}_{\mathrm{q}}^{j_1}(-\hat{x})\hat{\varphi}_{\mathrm{q}}^{j_2}(-\omega), \hat{\phi}_{\mathrm{u}}^{i_1}\hat{\varphi}_{\mathrm{u}}^{i_2}\hat{\mathcal{A}}_{\mathrm{q}}^{i_1}\hat{U} \right\rangle ;\end{aligned} \tag{8.26}$$

$$\begin{aligned}-\left\langle \hat{\phi}_{\mathrm{u}}^{j_1}(-\hat{x})\hat{\varphi}_{\mathrm{u}}^{j_2}(-\omega), \hat{\mathcal{A}}_{\mathrm{q}}^{j_1}\hat{u}_\chi \right\rangle = &- \left\langle \hat{\phi}_{\mathrm{u}}^{j_1}(-\hat{x})\hat{\varphi}_{\mathrm{u}}^{j_2}(-\omega), \hat{f}_\chi \hat{\mathcal{A}}_{\mathrm{q}}^{j_1}\hat{U} \right\rangle \\ &- \sum_{i_1,i_2} \mathsf{q}^{i_1 i_2} \left\langle \hat{\phi}_{\mathrm{u}}^{j_1}(-\hat{x})\hat{\varphi}_{\mathrm{u}}^{\mathrm{j}_2}(-\omega), \hat{\phi}_{\mathrm{q}}^{i_1}\hat{\varphi}_{\mathrm{q}}^{i_2}\hat{\mathcal{A}}_{\mathrm{q}}^{j_1}\hat{U} \right\rangle \\ &+ \sum_{i_1,i_2} \mathsf{u}^{i_1 i_2} \left\langle \hat{\phi}_{\mathrm{u}}^{j_1}(-\hat{x})\hat{\varphi}_{\mathrm{u}}^{j_2}(-\omega), \hat{\phi}_{\mathrm{u}}^{i_1}\hat{\varphi}_{\mathrm{u}}^{i_2}\hat{\mathcal{A}}_{\mathrm{q}}^{j_1}\hat{\mathcal{A}}_{\mathrm{q}}^{i_1}\hat{U} \right\rangle .\end{aligned} \tag{8.27}$$

8.1.5 Initial conditions

In (8.26) and (8.27) the initial conditions were neglected. To take them into account the Galerkin BIE (8.26) has to be extended on the right-hand side by, cf. [Bon95a],

$$\left\langle \hat{\phi}_{\mathrm{q}}^{j_1}(-\hat{x})\hat{\varphi}_{\mathrm{q}}^{j_2}(-\omega), \hat{u}_\chi^0 \hat{V} \right\rangle + \left\langle \hat{\phi}_{\mathrm{q}}^{j_1}(-\hat{x})\hat{\varphi}_{\mathrm{q}}^{j_2}(-\omega), \hat{v}_\chi^0 \hat{U} \right\rangle . \tag{8.28}$$

The additional terms for the second BIE (8.27) are

$$\left\langle \hat{\phi}_{\mathrm{m}}^{j_1}(-\hat{x})\hat{\varphi}_{\mathrm{m}}^{j_2}(-\omega), \hat{u}_\chi^0 \hat{\mathcal{A}}_{\mathrm{q}}^{j}\hat{V} \right\rangle + \left\langle \hat{\phi}_{\mathrm{q}}^{j_1}(-\hat{x})\hat{\varphi}_{\mathrm{q}}^{j_2}(-\omega), \hat{v}_\chi^0 \hat{U} \right\rangle . \tag{8.29}$$

with $\hat{u}_\chi^0$, $\hat{v}_\chi^o$ as the transformed initial values for u and $v = \partial_t u$ in the domain Ω at time $t = 0$. $V = \partial_t U \overset{\mathcal{F}}{\leftrightarrow} \hat{V} = i\omega\hat{U}$ is the time derivative of the fundamental solution U. These volume integrals might be evaluated directly, i.e. by a discretization of the domain with volume cells as explained in Chapter 10 for nonlinear problems. Alternatively, several authors have proposed special routines to transfer the volume integrals by a second Green's identity to the boundaries (double or multiple reciprocity BEM).

8.1.6 Transient plate problems

Thin plates

To obtain the dynamic analogue of the theory presented in Chapter 7.1, an inertia term and eventually a damping term has to be added to (7.3) :

$$\begin{aligned} (D\Delta\Delta + c\partial_t + h\rho\,\partial_{tt})\,w &= f \qquad (8.30) \\ \overset{\mathcal{F}}{\leftrightarrow} \quad \left(D(\hat{x}_1^2 + \hat{x}_2^2)^2 + ic\omega - h\rho\,\omega^2\right)\hat{w} &= \hat{f}, \end{aligned}$$

with ρ, c as the mass density and the viscous damping coefficient, respectively. To the authors knowledge, there are no transient fundamental solutions for the original space. The Fourier fundamental solution is

$$\hat{W}(\hat{x}, \omega) = \frac{1}{D(\hat{x}_1^2 + \hat{x}_2^2)^2 + ic\omega - h\rho\,\omega^2}. \qquad (8.31)$$

The boundary differential operators of the static problem are also valid for the dynamic case. Hence the Galerkin BIE is in analogy to (7.30)

$$\begin{aligned} \left\langle \hat{\phi}^{j_1}(-\hat{x})\hat{\varphi}^{j_2}(-\omega), \mathcal{B}\,\hat{w}_\chi \right\rangle &= \left\langle \hat{\phi}^{j_1}(-\hat{x})\hat{\varphi}^{j_2}(-\omega), \hat{f}_\chi \hat{\mathcal{B}}\,\hat{W} \right\rangle \qquad (8.32) \\ &+ \sum_{i_1, i_2} \hat{\mathrm{u}}^{i_1 i_2} \left\langle \hat{\phi}^{j_1}(-\hat{x})\hat{\varphi}^{j_2}(-\omega), \hat{\phi}^{i_1}\hat{\varphi}^{i_2}\hat{\mathcal{A}}\hat{W} \right\rangle. \end{aligned}$$

8.1.7 Dynamic analysis for bended and stressed plates

The consideration of in-plane forces leads to the differential equation

$$\begin{aligned} (D\Delta\Delta + c\partial_t + h\rho\,\partial_{tt} + N_{kl}\partial_{kl})\,w &= f \qquad (8.33) \\ \overset{\mathcal{F}}{\leftrightarrow} \quad \left(D(\hat{x}_1^2 + \hat{x}_2^2)^2 + ic\omega - h\rho\,\omega^2 - N_{kl}\hat{x}_k\hat{x}_l\right)\hat{w} &= \hat{f}, \end{aligned}$$

where N_{kl} are the known constant in-plane compressive forces. Hence, the fundamental solution for dynamic stability analysis is

$$\hat{W} = \frac{1}{D(\hat{x}_1^2 + \hat{x}_2^2)^2 + ic\omega - h\rho\,\omega^2 - N_{kl}\hat{x}_k\hat{x}_l}. \qquad (8.34)$$

By setting the right-hand side to zero, the dynamic stability of thin plates can be investigated.

For the particular case of isotropic in-plane forces $N_{kl} = \pm N$ (compressive/tensile) and vanishing damping, [Bes91] cites a fundamental solution in the frequency domain :

$$W = -\frac{1}{\alpha^2 + \gamma^2}\frac{i}{4}\left[\mathrm{H}_0^1(\alpha|x|) - \mathrm{H}_0^1(i\gamma|x|)\right], \qquad (8.35)$$

with $\zeta = N/D, \beta^4 = h\rho\omega^2/D$, and

$$\alpha^2 = \frac{1}{2}\left[\sqrt{\zeta^4 + 4\beta^4} \pm \zeta^2\right]; \qquad \gamma^2 = \frac{1}{2}\left[\sqrt{\zeta^4 + 4\beta^4} \mp \zeta^2\right].$$

8.1.8 Dynamic poroelasticity

The differential equation and its fundamental solution

The linear elastic model presented in Section 8.1.4 can be extended to describe the behaviour of a porous, fluid-saturated medium[2]. In addition to the Lamé coefficients λ, μ, two poroelastic coefficients have to be defined [Kyt96]:

$$R = \frac{2}{9}\gamma^2 B^2 \frac{(1-2\bar{\nu})(1+\bar{\nu}_\mathrm{u})^2}{(1-2\bar{\nu}_\mathrm{u})(\bar{\nu}_\mathrm{u}-\bar{\nu})}, \qquad Q = \frac{2}{3}\mu\gamma B \frac{1+\bar{\nu}_\mathrm{u}}{1-2\bar{\nu}_\mathrm{u}} - R, \tag{8.36}$$

where $\gamma, \bar{\nu}_\mathrm{u}, B$ denote the porosity, the undrained Poisson's ratio, and the Skempton pore pressure coefficient[3], respectively. Then, the stress-strain relations are

$$\begin{aligned} \sigma^\mathrm{s}_{kl} &= 2\mu\varepsilon^\mathrm{s}_{kl} + \left(\lambda + \frac{Q^2}{R}\right)\delta_{kl}\varepsilon^\mathrm{s}_{mm} + Q\delta_{kl}\varepsilon^\mathrm{f}_{mm}; \\ \sigma^\mathrm{f} &= Q\varepsilon^\mathrm{s}_{mm} + R\varepsilon^\mathrm{f}_{mm}. \end{aligned} \tag{8.37}$$

The superscripts s, f refer to the elastic solid and fluid parts. The fluid pore pressure is related to the fluid stress by

$$p = -\frac{1}{\gamma}\sigma^\mathrm{f}. \tag{8.38}$$

By assuming linear compatibility (small displacements of the solid u^s and of the fluid u^f)

$$\varepsilon^{\mathrm{s,f}}_{kl} = \frac{1}{2}\left(\partial_l u^{\mathrm{s,f}}_k + \partial_k u^{\mathrm{s,f}}_l\right), \tag{8.39}$$

the differential equations for the solid part

$$\begin{aligned} \left(\lambda + \mu + \frac{Q^2}{R}\right)\partial_{lk}u^\mathrm{s}_l + \mu\partial_{ll}u^\mathrm{s}_k + Q\partial_{lk}u^\mathrm{f}_l - F\left(\partial_t u^\mathrm{s}_k - \partial_t u^\mathrm{f}_k\right) \\ -\rho_{11}\partial_{tt}u^\mathrm{s}_k - \rho_{12}\partial_{tt}u^\mathrm{f}_k = -f^\mathrm{s}_k, \end{aligned} \tag{8.40}$$

and for the fluid part

$$Q\partial_{lk}u^\mathrm{s}_l + R\partial_{lk}u^\mathrm{f}_l + F\left(\partial_t u^\mathrm{s}_k - \partial_t u^\mathrm{f}_k\right) - \rho_{21}\partial_{tt}u^\mathrm{s}_k - \rho_{22}\partial_{tt}u^\mathrm{f}_k = -f^\mathrm{f}_k \tag{8.41}$$

[2] M.A. Biot: Theory of propagation of elastic waves in a fluid-saturated porous solid, Part I: Low-frequency range, *J. Acoust. Soc. Amer.* **28** (1956), 168–178; Part II. ibid **28** (1956), 179–191.

[3] A.W. Skempton: The pore pressure coefficients A and B. *Geotechnique* **4** (1954), 143–147.

are obtained. F is the viscous force generated by the relative movement between the solid and the fluid and represents viscous damping. ρ_{11}, ρ_{22} are the effective mass densities of the solid and the fluid, respectively, and $\rho_{12} = \rho_{21}$ is a dynamic coupling parameter. The transformed system of differential equations is

$$[\hat{\mathcal{P}}_1 + \hat{\mathcal{P}}_2](\hat{x}, \omega)\hat{u} = -\hat{f}, \tag{8.42}$$

with the transformed differential operators ($C_1 = \lambda + \mu + Q^2/R$)

$$\hat{\mathcal{P}}_1 = \begin{bmatrix} -C_1\hat{x}_1\hat{x}_1 & -C_1\hat{x}_1\hat{x}_2 & -C_1\hat{x}_1\hat{x}_3 & -Q\hat{x}_1\hat{x}_1 & -Q\hat{x}_1\hat{x}_2 & -Q\hat{x}_1\hat{x}_3 \\ -C_1\hat{x}_2\hat{x}_1 & -C_1\hat{x}_2\hat{x}_2 & -C_1\hat{x}_2\hat{x}_3 & -Q\hat{x}_2\hat{x}_1 & -Q\hat{x}_2\hat{x}_2 & -Q\hat{x}_2\hat{x}_3 \\ -C_1\hat{x}_3\hat{x}_1 & -C_1\hat{x}_3\hat{x}_2 & -C_1\hat{x}_3\hat{x}_3 & -Q\hat{x}_3\hat{x}_1 & -Q\hat{x}_3\hat{x}_2 & -Q\hat{x}_3\hat{x}_3 \\ -Q\hat{x}_1\hat{x}_1 & -Q\hat{x}_1\hat{x}_2 & -Q\hat{x}_1\hat{x}_3 & -R\hat{x}_1\hat{x}_1 & -R\hat{x}_1\hat{x}_2 & -R\hat{x}_1\hat{x}_3 \\ -Q\hat{x}_2\hat{x}_1 & -Q\hat{x}_2\hat{x}_2 & -Q\hat{x}_2\hat{x}_3 & -R\hat{x}_2\hat{x}_1 & -R\hat{x}_2\hat{x}_2 & -R\hat{x}_2\hat{x}_3 \\ -Q\hat{x}_3\hat{x}_1 & -Q\hat{x}_3\hat{x}_2 & -Q\hat{x}_3\hat{x}_3 & -R\hat{x}_3\hat{x}_1 & -R\hat{x}_3\hat{x}_2 & -R\hat{x}_3\hat{x}_3 \end{bmatrix};$$

$$\hat{\mathcal{P}}_2 = \begin{bmatrix} A_1 & 0 & 0 & A_2 & 0 & 0 \\ 0 & A_1 & 0 & 0 & A_2 & 0 \\ 0 & 0 & A_1 & 0 & 0 & A_2 \\ A_2 & 0 & 0 & A_3 & 0 & 0 \\ 0 & A_2 & 0 & 0 & A_3 & 0 \\ 0 & 0 & A_2 & 0 & 0 & A_3 \end{bmatrix}, \quad \begin{aligned} A_1 &= -\mu|x|^2 - iF\omega + \rho_{11}\omega^2; \\ A_2 &= +iF\omega + \rho_{12}\omega^2; \\ A_3 &= -iF\omega + \rho_{22}\omega^2, \end{aligned}$$

and with the displacements and volume forces

$$\begin{aligned} \hat{u}^T &= (\ \hat{u}_1^{\mathrm{s}}\ \ \hat{u}_2^{\mathrm{s}}\ \ \hat{u}_3^{\mathrm{s}}\ \ \hat{u}_1^{\mathrm{f}}\ \ \hat{u}_2^{\mathrm{f}}\ \ \hat{u}_3^{\mathrm{f}}\ ); \\ \hat{f}^T &= (\ \hat{f}_1^{\mathrm{s}}\ \ \hat{f}_2^{\mathrm{s}}\ \ \hat{f}_3^{\mathrm{s}}\ \ \hat{f}_1^{\mathrm{f}}\ \ \hat{f}_2^{\mathrm{f}}\ \ \hat{f}_3^{\mathrm{f}}\ ). \end{aligned}$$

Hence the Fourier fundamental solution is

$$\hat{U} = [\hat{\mathcal{P}}_1 + \hat{\mathcal{P}}_2]^{-1}. \tag{8.43}$$

For a BEM in the original space, see Wiebe [Wie93].

8.2 Stationary Problems

8.2.1 The stationarity assumption

Stationary problems are obtained by assuming that all quantities of the differential equation vary with $\mathrm{e}^{-i\omega_0 t}$, where ω_0 is a fixed value. Hence we have for a quantity a

$$a(x,t) = a(x,\omega_0)\,\mathrm{e}^{-i\omega_0 t} \quad \overset{\mathcal{F}}{\leftrightarrow} \quad \hat{a}(\hat{x},\omega) = \hat{a}(\hat{x},\omega_0)\delta(\omega-\omega_0). \tag{8.44}$$

The time dependency is separated; the quasi-static problem for $a(x,\omega_0) \overset{\mathcal{F}}{\leftrightarrow} a(\hat{x},\omega_0)$ can be solved first before reintroducing the sinusoidal time behaviour. Thus in the following the time dependency is omitted in the presentation.

8.2.2 Scalar waves in isotropic media

The fundamental solution and the Galerkin BIE

The stationary scalar wave equation is obtained by replacing $\partial_{tt} \to -\omega_0^2$ and $\omega \to \omega_0$ in (8.14) which results in the Helmholtz equation

$$\left(\Delta + k_{\rm p}^2\right) u(x) = -f(x) \qquad \overset{\mathcal{F}}{\leftrightarrow} \qquad \left(k_{\rm p}^2 - |\hat{x}|^2\right) \hat{u}(\hat{x}) = -\hat{f}(\hat{x}), \qquad (8.45)$$

with the given wave number $k_{\rm p} = \omega_0^2/c_{\rm p}^2$. The fundamental solutions are, cf. [Bon95a] for the original space,

$$\text{for } \mathbb{R}^2: \qquad U(x) = \frac{i}{4}\mathrm{H}_0^1(k_{\rm p}|x|) \ \overset{\mathcal{F}}{\leftrightarrow}\ \hat{U}(\hat{x}) = \frac{1}{k_{\rm p}^2 - |\hat{x}|^2}; \qquad (8.46)$$

$$\text{for } \mathbb{R}^3: \qquad U(x) = \frac{1}{4\pi|x|}\,\mathrm{e}^{ik_{\rm p}|x|} \ \overset{\mathcal{F}}{\leftrightarrow}\ \hat{U}(\hat{x}) = \frac{1}{k_{\rm p}^2 - |\hat{x}|^2}. \qquad (8.47)$$

H_0^1 is the zero order Hankel function of the first kind. The boundary differential operator is identical to that for the Poisson equation

$$\mathcal{A}_{\rm q}^i = -\nu^i \cdot \nabla \qquad \overset{\mathcal{F}}{\leftrightarrow} \qquad \hat{\mathcal{A}}_{\rm q}^i = -i\nu^i \cdot \hat{x}. \qquad (8.48)$$

The Galerkin BIE is similar to that of the Poisson equation, compare (5.32),

$$\begin{aligned}\left\langle \hat{\phi}_{\rm q}^j(-\hat{x}), \hat{u}_\chi \right\rangle &= \left\langle \hat{\phi}_{\rm q}^j(-\hat{x}), \hat{f}_\chi \hat{U} \right\rangle \qquad (8.49)\\ &- \sum_i^{N_{\rm q}} \mathsf{q}^i \left\langle \hat{\phi}_{\rm q}^j(-\hat{x}), \hat{\phi}_{\rm q}^i \hat{U} \right\rangle + \sum_i^{N_{\rm u}} \mathsf{u}^i \left\langle \hat{\phi}_{\rm q}^j(-\hat{x}), \hat{\phi}_{\rm u}^i \hat{\mathcal{A}}_{\rm q}^i \hat{U} \right\rangle;\\ \left\langle \hat{\phi}_{\rm u}^j(-\hat{x}), \hat{\mathcal{A}}_{\rm q}^j \hat{u}_\chi \right\rangle &= \left\langle \hat{\phi}_{\rm u}^j(-\hat{x}), \hat{f}_\chi \hat{\mathcal{A}}_{\rm q}^j \hat{U} \right\rangle \qquad (8.50)\\ &+ \sum_i^{N_{\rm q}} \mathsf{q}^i \left\langle \hat{\phi}_{\rm u}^j(-\hat{x}), \hat{\phi}_{\rm q}^i \hat{\mathcal{A}}_{\rm q}^j \hat{U} \right\rangle - \sum_i^{N_{\rm u}} \mathsf{u}^i \left\langle \hat{\phi}_{\rm u}^j(-\hat{x}), \hat{\phi}_{\rm u}^i \hat{\mathcal{A}}_{\rm q}^j \mathcal{A}_{\rm q}^i \hat{U} \right\rangle.\end{aligned}$$

The interior Dirichlet problem

The interior Dirichlet problem for the Helmholtz equation is

$$\begin{aligned}\left(\Delta + k_{\rm p}^2\right) u(x) &= -f(x), \qquad x \in \Omega; \qquad (8.51)\\ u = u_\Gamma &= 0, \qquad x \in \Gamma.\end{aligned}$$

Fig.8.3 shows the amplitude of the velocity potential u for $\omega = 10$ [rad/s] caused by a single source in the center of the square $\Omega = [0,1] \times [0,1]$. The

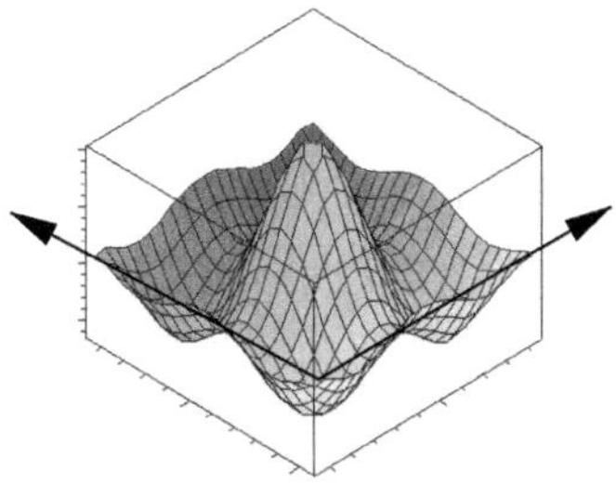

Figure 8.3: *The velocity potential u due to a single centered source with* $\omega_0 = 10$ *[rad/s]*

boundaries are discretized by four constant trial functions for each side which is not really sufficient for the accurate fulfillment of the imposed boundary conditions.

Sommerfeld's radiation condition

For exterior problems (e.g. holes in an infinite domain) an additional condition, Sommerfeld's radiation condition, is required to get unique results. It distinguishes between incoming and outgoing waves. We have for the $\mathbb{R}^n$ [DL00]

$$\begin{aligned} \text{incoming}: \quad & \left(\partial_{|x|} - ik_{\mathrm{p}}\right) u(x) = \mathcal{O}\left(\frac{1}{|x|^{(n+1)/2}}\right). \\ \text{outgoing}: \quad & \left(\partial_{|x|} + ik_{\mathrm{p}}\right) u(x) = \mathcal{O}\left(\frac{1}{|x|^{(n+1)/2}}\right). \end{aligned} \tag{8.52}$$

$\mathcal{O}(.)$ is the Landau symbol.

8.2.3 Elastic waves in isotropic media

The stationary fundamental solution

By substituting $\omega = \omega_0$ in the transient fundamental solution (8.23) we obtain

$$\hat{U} = \begin{bmatrix} c_s^2\hat{x}^2 + c_1^2\hat{x}_1^2 - \omega_0^2 & c_1^2\hat{x}_1\hat{x}_2 & c_1^2\hat{x}_1\hat{x}_3 \\ c_1^2\hat{x}_2\hat{x}_1 & c_s^2\hat{x}^2 + c_1^2\hat{x}_2^2 - \omega_0^2 & c_1^2\hat{x}_2\hat{x}_3 \\ c_1^2\hat{x}_3\hat{x}_1 & c_1^2\hat{x}_3\hat{x}_2 & c_s^2\hat{x}^2 + c_1^2\hat{x}_3^2 - \omega_0^2 \end{bmatrix}^{-1} . \quad (8.53)$$

The corresponding expression for the fundamental solution in the original space can be found for example in [Bon95a]. The stationary problems are not elaborated further, the derivation of the stationary fundamental solutions from the transient fundamental solution is always trivial in Fourier space.

Chapter 9

Thermoelasticity

J.M.C. Duhamel

9.1 Coupled Thermoelasticity

9.1.1 Coupled anisotropic thermoelasticity

The differential equation and its fundamental solution

The theory of thermoelasticity takes into account the coupling between elastic stresses and temperature, cf. [BSS89, Now75, Kög00]. We present first the fully coupled theory where the interaction between temperature and mechanical forces is taken into account. The equations are linearized and the heat conduction is modelled by Fourier's law. Then, several simplified models under additional assumptions are investigated. For all these problems, the Fourier fundamental solution and the BIE are derived. Fundamental solutions in the original space are only available for very special cases.

The two fundamental axioms for thermoelasticity are the first

$$\partial_t(e+k) = \partial_t w - \partial_k q_k \tag{9.1}$$

and the second law of thermodynamics

$$\partial_t s - \frac{\partial_t w}{T} + \partial_k \frac{q_k}{T} \geq 0. \tag{9.2}$$

[1] The first work on a coupled theory for stresses caused by forces and thermal stresses is [Tim53] : J.M.C. Duhamel: Mémoir sur le calcul des actions moléculaires développées par les changements du température dans les corps solides. *Mém. acad. sci. savants étrangers* **5** (1838), 440–498.

e, k are the internal and the kinetic energy per unit volume and $\partial_t w$ is the heat source density. q_k, s, T denote the components of the heat flux vector, the entropy per unit volume, and the temperature, respectively. The simplest (and linearized) model for the heat conduction is Fourier's law

$$q_k = -K_{kl}\partial_l T, \tag{9.3}$$

where K_{kl} is the symmetric thermal conductivity tensor. According to the Duhamel-Neumann relation, the stress tensor σ is depending on the strain tensor ε and the temperature differences $\theta = T - T_0$ (with T_0 as the temperature at the natural state)

$$\sigma_{kl} = C_{klmn}\varepsilon_{mn} - \beta_{kl}\theta. \tag{9.4}$$

The entropy s is related to ε and θ by

$$s = \beta_{kl}\varepsilon_{kl} + \frac{c_\varepsilon}{T_0}\theta. \tag{9.5}$$

C_{klmn} is the elasticity tensor measured at constant temperature and c_ε denotes the specific heat measured at constant strain. β_{kl} accounts for the interaction between strain and temperature. The dynamic equilibium is

$$\partial_l\sigma_{kl} - \rho\partial_{tt}u_k = -f_k. \tag{9.6}$$

ρ is the mass density and u, f denote the displacements and volume forces, respectively. The final system of differential equations describing the coupled thermoelasticity consists of the linearized dynamic equilibrium and the linearized heat conduction equation

$$\begin{aligned} C_{klmn}\partial_{nl}u_m - \rho\partial_{tt}u_k - \beta_{kl}\partial_l\theta &= -f_k; \\ K_{kl}\partial_{kl}\theta - c_\varepsilon\partial_t\theta - T_0\beta_{kl}\partial_t\partial_l u_k &= -\partial_t w. \end{aligned} \tag{9.7}$$

The transform of (9.7) leads to

$$\hat{\mathcal{P}}(\hat{x}, \omega)u = f, \tag{9.8}$$

$$\hat{\mathcal{P}} = \begin{bmatrix} C_{1l1n}\hat{x}_n\hat{x}_l - \rho\omega^2 & C_{1l2n}\hat{x}_n\hat{x}_l & C_{1l3n}\hat{x}_n\hat{x}_l & i\beta_{1l}\hat{x}_l \\ C_{2l1n}\hat{x}_n\hat{x}_l & C_{2l2n}\hat{x}_n\hat{x}_l - \rho\omega^2 & C_{2l3n}\hat{x}_n\hat{x}_l & i\beta_{2l}\hat{x}_l \\ C_{3l1n}\hat{x}_n\hat{x}_l & C_{3l2n}\hat{x}_n\hat{x}_l & C_{3l3n}\hat{x}_n\hat{x}_l - \rho\omega^2 & i\beta_{3l}\hat{x}_l \\ -T_0\beta_{1l}\omega\hat{x}_l & -T_0\beta_{2l}\omega\hat{x}_l & -T_0\beta_{3l}\omega\hat{x}_l & K_{kl}\hat{x}_k\hat{x}_l + i\omega c_\varepsilon \end{bmatrix}$$

and

$$u^T = (\ u_1 \quad u_2 \quad u_3 \quad \theta\ ); \qquad f^T = (\ f_1 \quad f_2 \quad f_3 \quad -i\omega w\ ).$$

The Fourier fundamental solution for coupled thermoelasticity in arbitrary anisotopic media is obtained by inversion

$$\hat{U} = \hat{\mathcal{P}}^{-1}(\hat{x}, \omega). \tag{9.9}$$

The boundary operator

The first quantity needed for the BEM algorithm is the traction t at the i-th boundary panel with the normal ν^i

$$t_k^i = C_{klmn}\partial_n\nu_l^i u_m - \beta_{kl}\nu_l^i\theta \quad \overset{\mathcal{F}}{\leftrightarrow} \quad \hat{t}_k^i = iC_{klmn}\hat{x}_n\nu_l^i\hat{u}_m - \beta_{kl}\nu_l^i\hat{\theta}. \tag{9.10}$$

As second term we need the normal heat flux

$$q_\nu^i = -K_{kl}\nu_k^i\partial_l\theta \quad \overset{\mathcal{F}}{\leftrightarrow} \quad \hat{q}_\nu^i = -iK_{kl}\nu_k^i\hat{x}_l\hat{\theta}. \tag{9.11}$$

These relations can be converted into operator notation, cf. [BSS89],

$$t^i = \mathcal{A}_{\mathrm{t}}^i u \quad \overset{\mathcal{F}}{\leftrightarrow} \quad \hat{t}^i = \hat{\mathcal{A}}_{\mathrm{t}}^i\hat{u}, \tag{9.12}$$

with u from (9.8) and the boundary operator

$$\begin{aligned}
\mathcal{A}_{\mathrm{t}}^i &= \begin{bmatrix} C_{1l1n}\nu_l^i\partial_n & C_{1l2n}\nu_l^i\partial_n & C_{1l3n}\nu_l^i\partial_n & -\beta_{1l}\nu_l^i \\ C_{2l1n}\nu_l^i\partial_n & C_{2l2n}\nu_l^i\partial_n & C_{2l3n}\nu_l^i\partial_n & -\beta_{2l}\nu_l^i \\ C_{3l1n}\nu_l^i\partial_n & C_{3l2n}\nu_l^i\partial_n & C_{3l3n}\nu_l^i\partial_n & -\beta_{3l}\nu_l^i \\ 0 & 0 & 0 & -K_{kl}\nu_k^i\partial_l \end{bmatrix} \\
\overset{\mathcal{F}}{\leftrightarrow} \quad \hat{\mathcal{A}}_{\mathrm{t}}^i &= \begin{bmatrix} iC_{1l1n}\nu_l^i\hat{x}_n & iC_{1l2n}\nu_l^i\hat{x}_n & iC_{1l3n}\nu_l^i\hat{x}_n & -\beta_{1l}\nu_l^i \\ iC_{2l1n}\nu_l^i\hat{x}_n & iC_{2l2n}\nu_l^i\hat{x}_n & iC_{2l3n}\nu_l^i\hat{x}_n & -\beta_{2l}\nu_l^i \\ iC_{3l1n}\nu_l^i\hat{x}_n & iC_{3l2n}\nu_l^i\hat{x}_n & iC_{3l3n}\nu_l^i\hat{x}_n & -\beta_{3l}\nu_l^i \\ 0 & 0 & 0 & -K_{kl}\nu_k^i\hat{x}_l \end{bmatrix}
\end{aligned} \tag{9.13}$$

and

$$(t^i)^T = (\ t_1^i \quad t_2^i \quad t_3^i \quad q_\nu^i\ ) \quad \overset{\mathcal{F}}{\leftrightarrow} \quad (\hat{t}^i)^T = (\ \hat{t}_1^i \quad \hat{t}_2^i \quad \hat{t}_3^i \quad \hat{q}_\nu^i.\ )$$

The Galerkin BIE

The Galerkin BIE can be adopted from elastodynamics, i.e. from (8.26) and (8.27), by inserting the fundamental solution (9.9) and the boundary operator (9.13). This leads to the displacement Galerkin BIE

$$\begin{aligned}
\left\langle \hat{\phi}_{\mathrm{t}}^{j_1}(-\hat{x})\hat{\varphi}_{\mathrm{t}}^{j_2}(-\omega), \hat{u}_\chi \right\rangle &= \left\langle \hat{\phi}_{\mathrm{t}}^{j_1}(-\hat{x})\hat{\varphi}_{\mathrm{t}}^{j_2}(-\omega), \hat{f}_\chi\hat{U} \right\rangle \\
&+ \sum_{i_1,i_2} \mathrm{t}^{i_1i_2} \left\langle \hat{\phi}_{\mathrm{t}}^{j_1}(-\hat{x})\hat{\varphi}_{\mathrm{t}}^{j_2}(-\omega), \hat{\phi}_{\mathrm{t}}^{i_1}\hat{\varphi}_{\mathrm{t}}^{i_2}\hat{U} \right\rangle \\
&- \sum_{i_1,i_2} \mathrm{u}^{i_1i_2} \left\langle \hat{\phi}_{\mathrm{t}}^{j_1}(-\hat{x})\hat{\varphi}_{\mathrm{t}}^{j_2}(-\omega), \hat{\phi}_{\mathrm{u}}^{i_1}\hat{\varphi}_{\mathrm{u}}^{i_2}\hat{\mathcal{A}}_{\mathrm{t}}^{i_1}\hat{U} \right\rangle;
\end{aligned} \tag{9.14}$$

and the traction Galerkin BIE :

$$
\begin{aligned}
-\left\langle \hat{\phi}_{\mathrm{u}}^{j_1}(-\hat{x})\hat{\varphi}_{\mathrm{u}}^{j_2}(-\omega), \hat{\mathcal{A}}_{\mathrm{t}}^{j_1}\hat{u}_{\chi} \right\rangle = & -\left\langle \hat{\phi}_{\mathrm{u}}^{j_1}(-\hat{x})\hat{\varphi}_{\mathrm{u}}^{j_2}(-\omega), \hat{f}_{\chi}\hat{\mathcal{A}}_{\mathrm{t}}^{j_1}\hat{U} \right\rangle \\
& -\sum_{i_1,i_2} \mathrm{t}^{i_1 i_2} \left\langle \hat{\phi}_{\mathrm{u}}^{j_1}(-\hat{x})\hat{\varphi}_{\mathrm{u}}^{j_2}(-\omega), \hat{\phi}_{\mathrm{t}}^{i_1}\hat{\varphi}_{\mathrm{t}}^{i_2}\hat{\mathcal{A}}_{\mathrm{t}}^{j_1}\hat{U} \right\rangle \\
& +\sum_{i_1,i_2} \mathrm{u}^{i_1 i_2} \left\langle \hat{\phi}_{\mathrm{u}}^{j_1}(-\hat{x})\hat{\varphi}_{\mathrm{u}}^{j_2}(-\omega), \hat{\varphi}_{\mathrm{u}}^{i_1}\hat{\phi}_{\mathrm{u}}^{i_2}\hat{\mathcal{A}}_{\mathrm{t}}^{j_1}\hat{\mathcal{A}}_{\mathrm{t}}^{i_1}\hat{U} \right\rangle .
\end{aligned}
\tag{9.15}
$$

9.1.2 Coupled isotropic thermoelasticity

The differential equation and its fundamental solution

For isotropic media the elasticity components are

$$C_{klmn} = \lambda\delta_{kl}\delta_{mn} + \mu(\delta_{km}\delta_{ln} + \delta_{kn}\delta_{lm}), \tag{9.16}$$

and the isotropic thermal constants are, cf. [BSS89],

$$\beta_{kl} = \gamma_0\delta_{kl}, \qquad \lambda_{kl} = \lambda_0\delta_{kl}, \tag{9.17}$$

where γ_0 depends on the coefficient of linear thermal expansion α_t

$$\gamma_0 = (\mu + 3\lambda)\alpha_t.$$

The thermal diffusivity κ_0 is

$$\kappa_0 = \frac{\lambda_0}{c - \varepsilon}. \tag{9.18}$$

The differential equation (9.7) is then simplified to, cf. [BSS89],

$$
\begin{aligned}
\mu\partial_{ll}u_k + (\lambda + \mu)\partial_{kl}u_l - \rho\partial_{tt}u_k - \gamma_0\partial_k\theta &= -f_k \\
\partial_{kk}\theta - \frac{1}{\kappa_0}\partial_t\theta - \eta_0\partial_t\partial_k u_k &= -\frac{Q}{\kappa_0},
\end{aligned}
\tag{9.19}
$$

with $\eta_0 = \gamma_0 T_0/\lambda_0$ and $Q = w/c_\varepsilon$. The Fourier fundamental solution is hence

$$\hat{U} = \hat{\mathcal{P}}^{-1}(\hat{x}, \omega); \tag{9.20}$$

where the transformed differential operator is defined as

$$
\hat{\mathcal{P}} = \begin{bmatrix}
-c_1\hat{x}_1\hat{x}_1 - A_1 & -c_1\hat{x}_1\hat{x}_2 & -c_1\hat{x}_1\hat{x}_3 & -i\hat{x}_1\gamma_0 \\
-c_1\hat{x}_2\hat{x}_1 & -c_1\hat{x}_2\hat{x}_2 - A_1 & -c_1\hat{x}_2\hat{x}_3 & -i\hat{x}_2\gamma_0 \\
-c_1\hat{x}_3\hat{x}_1 & -c_1\hat{x}_3\hat{x}_2 & -c_1\hat{x}_3\hat{x}_3 - A_1 & -i\hat{x}_3\gamma_0 \\
\eta_0\hat{x}_1\omega & \eta_0\hat{x}_2\omega & \eta_0\hat{x}_3\omega & -|\hat{x}|^2 - i\omega/\kappa_0
\end{bmatrix}
$$

with $c_1 = \lambda + \mu$ and $A_1 = \mu|\hat{x}|^2 - \rho\omega^2$.

9.2 Simplified Thermoelastic Models

9.2.1 Thermal stresses

In cases where the deformation is mainly caused by thermal sources than by external mechanical forces, it can be assumed that the coupling term $T_0\beta_{kl}\partial_t\partial_l u_k$ is negligible compared to $c_\varepsilon\partial_t\theta$, cf. [Kög00]. This means that the thermoelastic dissipation has no significant influence on the thermal field. The second equation in (9.7) is then decoupled from the first

$$K_{kl}\partial_{kl}\theta - c_\varepsilon\partial_t\theta = -\partial_t w \tag{9.21}$$

and solved seperately. In a second step, the obtained temperature field θ is regarded as "body force" for the elastic fields in the first equation of (9.7). The fundamental solution is then

$$\hat{U} = \begin{bmatrix} C_{1l1n}\hat{x}_n\hat{x}_l - \rho\omega^2 & C_{1l2n}\hat{x}_n\hat{x}_l & C_{1l3n}\hat{x}_n\hat{x}_l & i\beta_{1l}\hat{x}_l \\ C_{2l1n}\hat{x}_n\hat{x}_l & C_{2l2n}\hat{x}_n\hat{x}_l - \rho\omega^2 & C_{2l3n}\hat{x}_n\hat{x}_l & i\beta_{2l}\hat{x}_l \\ C_{3l1n}\hat{x}_n\hat{x}_l & C_{3l2n}\hat{x}_n\hat{x}_l & C_{3l3n}\hat{x}_n\hat{x}_l - \rho\omega^2 & i\beta_{3l}\hat{x}_l \\ 0 & 0 & 0 & K_{kl}\hat{x}_k\hat{x}_l + i\omega c_\varepsilon \end{bmatrix}^{-1}$$

9.2.2 Coupled quasi-static thermoelasticity

If the external sources f, w vary slowly in time compared with the characteristic frequency c/L the inertia terms can be omitted (c is the speed of the elastic waves and L the characteristic length of the body). The coupled system of quasi-static differential equations is then

$$\begin{aligned} C_{klmn}\partial_{nl}u_m - \beta_{kl}\partial_l\theta &= -f_k; \\ K_{kl}\partial_{kl}\theta - c_\varepsilon\partial_t\theta - T_0\beta_{kl}\partial_t\partial_l u_k &= -\partial_t w. \end{aligned} \tag{9.22}$$

The Fourier fundamental solution is

$$\hat{U} = \begin{bmatrix} C_{1l1n}\hat{x}_n\hat{x}_l & C_{1l2n}\hat{x}_n\hat{x}_l & C_{1l3n}\hat{x}_n\hat{x}_l & i\beta_{1l}\hat{x}_l \\ C_{2l1n}\hat{x}_n\hat{x}_l & C_{2l2n}\hat{x}_n\hat{x}_l & C_{2l3n}\hat{x}_n\hat{x}_l & i\beta_{2l}\hat{x}_l \\ C_{3l1n}\hat{x}_n\hat{x}_l & C_{3l2n}\hat{x}_n\hat{x}_l & C_{3l3n}\hat{x}_n\hat{x}_l & i\beta_{3l}\hat{x}_l \\ -T_0\beta_{1l}\omega\hat{x}_l & -T_0\beta_{2l}\omega\hat{x}_l & -T_0\beta_{3l}\omega\hat{x}_l & K_{kl}\hat{x}_k\hat{x}_l + i\omega c_\varepsilon \end{bmatrix}^{-1}$$

9.2.3 Uncoupled quasi-static thermoelasticity

Combining the two approximations, i.e. neglecting the coupling term $T_0\beta_{kl}\partial_t\partial_l u_k$ and the inertia term $-\rho\partial_{tt}u_k$, the uncoupled quasi-static model is obtained. The differential equation is then

$$\begin{aligned} C_{klmn}\partial_{nl}u_m - \beta_{kl}\partial_l\theta &= -f_k; \\ K_{kl}\partial_{kl}\theta - c_\varepsilon\partial_t\theta &= -\partial_t w, \end{aligned} \tag{9.23}$$

and as fundamental solution we get

$$\hat{U} = \begin{bmatrix} C_{1l1n}\hat{x}_n\hat{x}_l & C_{1l2n}\hat{x}_n\hat{x}_l & C_{1l3n}\hat{x}_n\hat{x}_l & i\beta_{1l}\hat{x}_l \\ C_{2l1n}\hat{x}_n\hat{x}_l & C_{2l2n}\hat{x}_n\hat{x}_l & C_{2l3n}\hat{x}_n\hat{x}_l & i\beta_{2l}\hat{x}_l \\ C_{3l1n}\hat{x}_n\hat{x}_l & C_{3l2n}\hat{x}_n\hat{x}_l & C_{3l3n}\hat{x}_n\hat{x}_l & i\beta_{3l}\hat{x}_l \\ 0 & 0 & 0 & K_{kl}\hat{x}_k\hat{x}_l + i\omega c_\varepsilon \end{bmatrix}^{-1} \tag{9.24}$$

9.2.4 Stationary thermoelasticity

In the stationary state, all physical variables are independent of time, leading to the Poisson equation for heat conduction and to Navier's equation for elasticity with a body source term caused by the temperature

$$\begin{aligned} C_{klmn}\partial_{nl}u_m - \beta_{kl}\partial_l\theta &= -f_k; \\ K_{kl}\partial_{kl}\theta &= -\partial_t w. \end{aligned} \tag{9.25}$$

The corresponding fundamental solution is

$$\hat{U} = \begin{bmatrix} C_{1l1n}\hat{x}_n\hat{x}_l & C_{1l2n}\hat{x}_n\hat{x}_l & C_{1l3n}\hat{x}_n\hat{x}_l & i\beta_{1l}\hat{x}_l \\ C_{2l1n}\hat{x}_n\hat{x}_l & C_{2l2n}\hat{x}_n\hat{x}_l & C_{2l3n}\hat{x}_n\hat{x}_l & i\beta_{2l}\hat{x}_l \\ C_{3l1n}\hat{x}_n\hat{x}_l & C_{3l2n}\hat{x}_n\hat{x}_l & C_{3l3n}\hat{x}_n\hat{x}_l & i\beta_{3l}\hat{x}_l \\ 0 & 0 & 0 & K_{kl}\hat{x}_k\hat{x}_l \end{bmatrix}^{-1} \tag{9.26}$$

For some fundamental solutions in the original space see Balaš et al. [BSS89].

Chapter 10

Non-linearity

R. von Mises

10.1 Physical Non-linearity

10.1.1 Inelastic problems

The differential equation for inelastic materials

In the following, all physical terms are given in rate terms. Assuming linear strain-displacement relationship, the total strain rate can be decomposed into a linear $\dot{\varepsilon}^{\mathrm{L}}_{kl}$ and a non-linear part $\dot{\varepsilon}^{\mathrm{N}}_{kl}$, cf. [Dal93],

$$\dot{\varepsilon}_{kl} = \frac{1}{2}(\partial_l \dot{u}_k + \partial_k \dot{u}_l) = \dot{\varepsilon}^{\mathrm{L}}_{kl} + \dot{\varepsilon}^{\mathrm{N}}_{kl}. \tag{10.1}$$

The linear part is related by Hooke's law to the elastic stress rates through, compare (6.22),

$$\dot{\sigma}_{kl} = C_{klmn}\dot{\varepsilon}^{\mathrm{L}}_{mn}, \tag{10.2}$$

with $C_{klmn} = \lambda\,\delta_{kl}\delta_{mn} + 2\mu\,\delta_{km}\delta_{ln}$ in the isotropic case. The substitution of $\dot{\varepsilon}^{\mathrm{L}}$ by $\dot{\varepsilon} - \dot{\varepsilon}^{\mathrm{N}}$ yields

$$\dot{\sigma}_{kl} = C_{klmn}\dot{\varepsilon}_{mn} - \dot{\sigma}^{\mathrm{N}}_{kl}, \qquad \dot{\sigma}^{\mathrm{N}}_{kl} = C_{klmn}\dot{\varepsilon}^{\mathrm{N}}_{mn}. \tag{10.3}$$

The equilibrium equation is

$$\partial_l \dot{\sigma}_{kl} = -\dot{f}_k. \tag{10.4}$$

which leads after introduction of (10.1) to the differential equation

$$C_{klmn}\partial_n \partial_l \dot{u}_m = -\dot{f}_k + \partial_l \dot{\sigma}^{\mathrm{N}}_{kl}. \tag{10.5}$$

The left-hand side of (10.5) is identical to that obtained in the linear case that means all non-linear contributions can be regarded as fictitious volume forces. Similarly, we get fictitious tractions on the boundary

$$\dot{t}_k^{\mathrm{N}} = \nu_l \dot{\sigma}_{kl}^{\mathrm{N}}. \tag{10.6}$$

Hence, the non-linear problem is reformulated as a linear one where the non-linearity is put on the right-hand side and treated as initial stress and boundary tractions. The final solution is obtained by a standard iteration process, e.g. [Dal93]. Each step consists of a totally linear problem and is solved as discussed in the previous chapters.

The incremental boundary integral equations

Equation (10.5) can be solved by a BEM based on the linear fundamental solution U. Transferring the collocation approach proposed by Dallner [Dal93] to a Galerkin method, the following BIE can be defined in Fourier space

$$\left\langle \hat{\phi}_{\mathrm{t}}^j(-\hat{x}), \hat{\dot{u}}_\chi \right\rangle = \left\langle \hat{\phi}_{\mathrm{t}}^j(-\hat{x}), \hat{\dot{f}}_\chi \hat{U} \right\rangle + \left\langle \hat{\phi}_{\mathrm{t}}^j(-\hat{x}), \hat{\dot{\varepsilon}}^{\mathrm{N}} \hat{\mathcal{A}}_\sigma \hat{U} \right\rangle \tag{10.7}$$

$$+ \sum_{N_{\mathrm{t}}}^{i} \mathrm{t}^i \left\langle \hat{\phi}_{\mathrm{t}}^j(-\hat{x}), \hat{\phi}_{\mathrm{t}}^i \hat{U} \right\rangle - \sum_{i}^{N_{\mathrm{u}}} \mathrm{u}^i \left\langle \hat{\phi}_{\mathrm{t}}^j(-\hat{x}), \hat{\phi}_{\mathrm{u}}^i \hat{\mathcal{A}}_{\mathrm{t}}^i \hat{U} \right\rangle;$$

$$-\left\langle \hat{\phi}_{\mathrm{u}}^j(-\hat{x}), \hat{\mathcal{A}}_{\mathrm{t}}^j \hat{\dot{u}}_\chi \right\rangle = -\left\langle \hat{\phi}_{\mathrm{u}}^j(-\hat{x}), \hat{\dot{f}}_\chi \hat{\mathcal{A}}_{\mathrm{t}}^j \hat{U} \right\rangle - \left\langle \hat{\phi}_{\mathrm{u}}^j(-\hat{x}), \hat{\dot{\varepsilon}}^{\mathrm{N}} \hat{\mathcal{A}}_{\mathrm{t}}^j \hat{\mathcal{A}}_\sigma \hat{U} \right\rangle \tag{10.8}$$

$$- \sum_{i}^{N_{\mathrm{t}}} \mathrm{t}^i \left\langle \hat{\phi}_{\mathrm{u}}^j(-\hat{x}), \hat{\phi}_{\mathrm{t}}^i \hat{\mathcal{A}}_{\mathrm{t}}^j \hat{U} \right\rangle + \sum_{i}^{N_{\mathrm{u}}} \mathrm{u}^i \left\langle \hat{\phi}_{\mathrm{u}}^j(-\hat{x}), \hat{\phi}_{\mathrm{u}}^i \hat{\mathcal{A}}_{\mathrm{t}}^j \hat{\mathcal{A}}_{\mathrm{t}}^i \hat{U} \right\rangle.$$

The only difference to the linear case lies in the integral containing the non-linear strain $\hat{\dot{\varepsilon}}^{\mathrm{N}}$. U is the linear fundamental solution and $\mathcal{A}_{\mathrm{t}}^j$ is the corresponding boundary operator. $\mathcal{A}_\sigma$ is the stress differential operator

$$\mathcal{A}_\sigma = C_{klmn} \partial_n \quad \overset{\mathcal{F}}{\leftrightarrow} \quad \hat{\mathcal{A}}_\sigma = i C_{klmn} \hat{x}_n, \tag{10.9}$$

which generates the fundamental stress tensor Σ_{pkl} from the fundamental solution U_{pm} :

$$\Sigma_{pkl} = C_{klmn} \partial_n U_{pm} \quad \overset{\mathcal{F}}{\leftrightarrow} \quad \hat{\Sigma}_{pkl} = i C_{klmn} \hat{x}_n \hat{U}_{pm}. \tag{10.10}$$

The discretization of the locally plastifying zone

The discretization of the boundary quantities in (10.7) and (10.8) can be

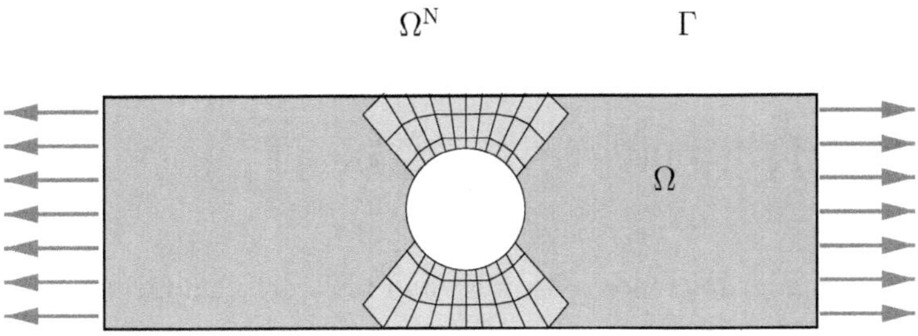

Figure 10.1: *The domain discretization of a locally plastifying region estimated a priori, cf. [Dal93]*

taken directly from the linear case. Required is still a partition of the plastifying zone Ω^{N} in to cells, see Fig.10.1. The non-linear strain rate $\dot{\varepsilon}^{\mathrm{N}}$ for the next increment in the iteration process has to be approximated in the inner cells by local volume trial functions (or alternatively by a dual reciprocity concept which uses global trial functions like radial basis functions, wavelets, etc., e.g. [Pol96]). This zone is a priori not known but can often be estimated roughly. The rest of the domain Ω remains undiscretized as long as volume forces f are zero.

The non-linear strain rate $\dot{\varepsilon}^{\mathrm{N}}_{kl}$ is approximated here by N_{e} standard polynomial trial functions

$$\dot{\varepsilon}^{\mathrm{N}}_{kl}(x) \approx \sum_{i}^{N_{\mathrm{e}}} \dot{\mathbf{e}}^{i}_{kl} \phi^{i}_{\mathrm{e}}(x) \qquad \overset{\mathcal{F}}{\leftrightarrow} \qquad \hat{\dot{\varepsilon}}^{\mathrm{N}}_{kl}(\hat{x}) \approx \sum_{i}^{N_{\mathrm{e}}} \dot{\mathbf{e}}^{i}_{kl} \hat{\phi}^{i}_{\mathrm{e}}(\hat{x}). \tag{10.11}$$

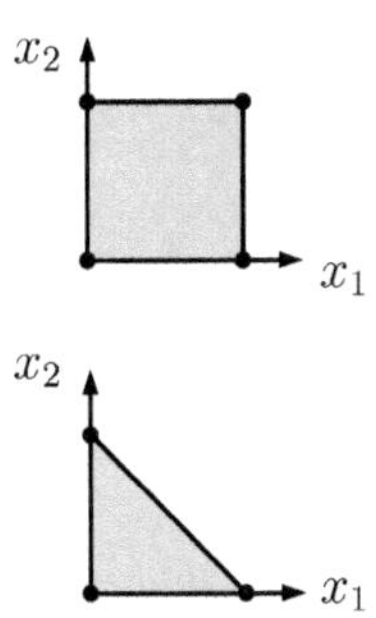

$$\phi^1_{\mathrm{e}} = (1 - x_1)(1 - x_2)\chi^0_{\mathrm{e}} \qquad \phi^2_{\mathrm{e}} = x_1(1 - x_2)\chi^0_{\mathrm{e}}$$

$$\phi^3_{\mathrm{e}} = (1 - x_1)x_2\chi^0_{\mathrm{e}} \qquad \phi^4_{\mathrm{e}} = x_1 x_2 \chi^0_{\mathrm{e}}$$

$$\phi^1_{\mathrm{e}} = 1 - x_1 - x_2\chi^0_{\mathrm{e}} \qquad \phi^2_{\mathrm{e}} = x_1\chi^0_{\mathrm{e}}$$

$$\phi^3_{\mathrm{e}} = x_2\chi^0_{\mathrm{e}}$$

Figure 10.2: *Linear trial functions for rectangular / triangular elements ($\mathbb{R}^2$)*

The cutoff distributions χ_{e}^0 for rectangular reference cells are for the $\mathbb{R}^n$

$$\chi_{\mathrm{e}}^0(x) = \prod_{k=1}^{n} \mathrm{H}(x_k)\mathrm{H}(1-x_k) \quad \overset{\mathcal{F}}{\leftrightarrow} \quad \hat{\chi}_{\mathrm{e}}^0(\hat{x}) = \prod_{k=1}^{n} \frac{i}{\hat{x}_k}\left(e^{-i\hat{x}_k} - 1\right), \tag{10.12}$$

and for triangular reference cells in $\mathbb{R}^2$ and in $\mathbb{R}^3$, cf. [Dud97b],

$$\begin{aligned}
\text{for } \mathbb{R}^2: \quad \chi_{\mathrm{e}}^0(x) &= \mathrm{H}(x_1)\mathrm{H}(x_2)\mathrm{H}(1-x_1-x_2) \qquad (10.13)\\
\overset{\mathcal{F}}{\leftrightarrow} \quad \hat{\chi}_{\mathrm{e}}^0(\hat{x}) &= \frac{1}{\hat{x}_1-\hat{x}_2}\left[\frac{1}{\hat{x}_1}\left(1-e^{-i\hat{x}_1}\right) - \frac{1}{\hat{x}_2}\left(1-e^{-i\hat{x}_2}\right)\right];\\
\text{for } \mathbb{R}^3: \quad \chi_{\mathrm{e}}^0(x) &= \mathrm{H}(x_1)\mathrm{H}(x_2)\mathrm{H}(x_3)\mathrm{H}(1-x_1-x_2-x_3)\\
\overset{\mathcal{F}}{\leftrightarrow} \quad \hat{\chi}_{\mathrm{e}}^0(\hat{x}) &= \frac{i(\hat{x}_1\hat{x}_2+\hat{x}_1\hat{x}_3-\hat{x}_1^2)(e^{-i\hat{x}_2}+e^{-i\hat{x}_3})}{\hat{x}_1\hat{x}_2\hat{x}_3(\hat{x}_1-\hat{x}_2)(\hat{x}_1-\hat{x}_3)}\\
&\quad - \frac{ie^{-i\hat{x}_1}+ie^{i(\hat{x}_1-\hat{x}_2-\hat{x}_3)}}{\hat{x}_1(\hat{x}_1-\hat{x}_2)(\hat{x}_1-\hat{x}_3)} + \frac{i+ie^{-i(\hat{x}_2+\hat{x}_3)}}{\hat{x}_1\hat{x}_2\hat{x}_3}.
\end{aligned}$$

Arbitrary trial functions ϕ_{e}^i of arbitrary polynomial degree defined on translated, rotated and dilated cells are obtained by the application of (3.19) and (3.20). Fig.10.2 shows the rectangular and triangular reference elements with linear trial functions for the $\mathbb{R}^2$. Fig.10.3 gives the same for three dimensions.

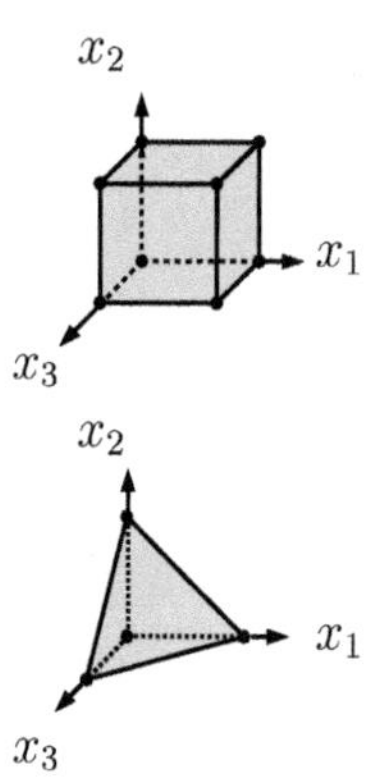

$\phi_{\mathrm{e}}^1 = (1-x_1)(1-x_2)(1-x_3)\chi^0$ $\quad \phi_{\mathrm{e}}^5 = x_1x_2(1-x_3)\chi^0$

$\phi_{\mathrm{e}}^2 = x_1(1-x_2)(1-x_3)\chi^0$ $\quad \phi_{\mathrm{e}}^6 = x_1(1-x_2)x_3\chi^0$

$\phi_{\mathrm{e}}^3 = (1-x_1)x_2(1-x_3)\chi^0$ $\quad \phi_{\mathrm{e}}^7 = (1-x_1)x_2x_3\chi^0$

$\phi_{\mathrm{e}}^4 = (1-x_1)(1-x_2)x_3\chi^0$ $\quad \phi_{\mathrm{e}}^8 = x_1x_2x_3\chi^0$

$\phi_{\mathrm{e}}^1 = 1-x_1-x_2-x_3\chi^0$ $\quad \phi_{\mathrm{e}}^3 = x_1(1-x_2)x_3\chi^0$

$\phi_{\mathrm{e}}^2 = (1-x_1)x_2x_3\chi^0$ $\quad \phi_{\mathrm{e}}^4 = x_1x_2(1-x_3)\chi^0$

Figure 10.3: *Linear trial functions for rectangular / triangular elements ($\mathbb{R}^3$)*

The additional BIE for the interior degrees of freedom

The discretized versions of (10.7) and (10.8) are now

$$\left\langle \hat{\phi}^j_{\mathrm{t}}(-\hat{x}), \hat{\dot{u}}_\chi \right\rangle = \left\langle \hat{\phi}^j_{\mathrm{t}}(-\hat{x}), \hat{\dot{f}}_\chi \hat{U} \right\rangle + \sum_i^{N_{\mathrm{e}}} \mathfrak{e}^i \left\langle \hat{\phi}^j_{\mathrm{t}}(-\hat{x}), \hat{\phi}^i_{\mathrm{e}} \hat{\mathcal{A}}_\sigma \hat{U} \right\rangle \tag{10.14}$$

$$+ \sum_{N_{\mathrm{t}}}^{i} \mathfrak{t}^i \left\langle \hat{\phi}^j_{\mathrm{t}}(-\hat{x}), \hat{\phi}^i_{\mathrm{t}} \hat{U} \right\rangle - \sum_i^{N_{\mathrm{u}}} \mathfrak{u}^i \left\langle \hat{\phi}^j_{\mathrm{t}}(-\hat{x}), \hat{\phi}^i_{\mathrm{u}} \hat{\mathcal{A}}^i_{\mathrm{t}} \hat{U} \right\rangle;$$

$$- \left\langle \hat{\phi}^j_{\mathrm{u}}(-\hat{x}), \hat{\mathcal{A}}^j_{\mathrm{t}} \hat{\dot{u}}_\chi \right\rangle = - \left\langle \hat{\phi}^j_{\mathrm{u}}(-\hat{x}), \hat{\dot{f}}_\chi \hat{\mathcal{A}}^j_{\mathrm{t}} \hat{U} \right\rangle - \sum_i^{N_{\mathrm{e}}} \mathfrak{e}^i \left\langle \hat{\phi}^j_{\mathrm{u}}(-\hat{x}), \hat{\phi}^i_{\mathrm{e}} \hat{\mathcal{A}}^j_{\mathrm{t}} \hat{\mathcal{A}}_\sigma \hat{U} \right\rangle$$

$$- \sum_i^{N_{\mathrm{t}}} \mathfrak{t}^i \left\langle \hat{\phi}^j_{\mathrm{u}}(-\hat{x}), \hat{\phi}^i_{\mathrm{t}} \hat{\mathcal{A}}^j_{\mathrm{t}} \hat{U} \right\rangle + \sum_i^{N_{\mathrm{u}}} \mathfrak{u}^i \left\langle \hat{\phi}^j_{\mathrm{u}}(-\hat{x}), \hat{\phi}^i_{\mathrm{u}} \hat{\mathcal{A}}^j_{\mathrm{t}} \hat{\mathcal{A}}^i_{\mathrm{t}} \hat{U} \right\rangle.$$

For the additional degrees of freedom, the coefficients $\mathfrak{e}^i$ of the interior strain rates $\dot{\varepsilon}^{\mathrm{N}}_{kl}$, the following BIEs are required, cf. [Dal93],

$$\left\langle \hat{\phi}^j_{\mathrm{e}}(-\hat{x}), \hat{\mathcal{A}}_\varepsilon \hat{\dot{u}} \right\rangle = \left\langle \hat{\phi}^j_{\mathrm{e}}(-\hat{x}), \hat{\dot{f}}_\chi \hat{\mathcal{A}}_\varepsilon \hat{U} \right\rangle + \sum_i^{N_{\mathrm{e}}} \mathfrak{e}^i \left\langle \hat{\phi}^j_{\mathrm{e}}(-\hat{x}), \hat{\phi}^i_{\mathrm{e}} \hat{\mathcal{A}}_\varepsilon \hat{\mathcal{A}}_\sigma \hat{U} \right\rangle \tag{10.15}$$

$$+ \sum_{N_{\mathrm{t}}}^{i} \mathfrak{t}^i \left\langle \hat{\phi}^j_{\mathrm{e}}(-\hat{x}), \hat{\phi}^i_{\mathrm{t}} \hat{\mathcal{A}}_\varepsilon \hat{U} \right\rangle - \sum_i^{N_{\mathrm{u}}} \mathfrak{u}^i \left\langle \hat{\phi}^j_{\mathrm{e}}(-\hat{x}), \hat{\phi}^i_{\mathrm{u}} \hat{\mathcal{A}}_\varepsilon \hat{\mathcal{A}}^i_{\mathrm{t}} \hat{U} \right\rangle,$$

with

$$\mathcal{A}_\varepsilon U = \frac{1}{2}(\partial_l U_{km} + \partial_k U_{lm}) \quad \overset{\mathcal{F}}{\leftrightarrow} \quad \hat{\mathcal{A}}_\varepsilon \hat{U} = \frac{i}{2}(\hat{x}_l \hat{U}_{km} + \hat{x}_k \hat{U}_{lm}). \tag{10.16}$$

The algorithm for the solution is not presented here, it can be taken for example from Dallner [Dal93]. As for the linear case, the Fourier BEM approach enables the solution of non-linear problems where the fundamental solution is not known.

10.2 Geometrical Non-linearity

10.2.1 Large deflection of thin elastic plates

The von Kármán equation

As an example of a BEM for geometrical non-linear problems, the moderately large deflection of thin elastic plates is considered. Here, it is assumed that the strains in the mid-plane (x_1, x_2) of the plate are given by, cf. [Kyt96, WJT00],

$$\begin{aligned} \varepsilon_{kl} &= \varepsilon_{kl}^{\mathrm{L}} + \varepsilon_{kl}^{\mathrm{N}}, & k,l = 1,2; \\ \text{with} \quad \varepsilon_{kl}^{\mathrm{L}} &= \frac{1}{2}(\partial_l u_k + \partial_k u_l), & \varepsilon_{kl}^{\mathrm{N}} = \frac{1}{2}\partial_k w \partial_l w. \end{aligned} \tag{10.17}$$

This means that all non-linear terms in the non-linear compatibility equation of elastic continua

$$\varepsilon_{kl} = \frac{1}{2}(\partial_l u_k + \partial_k u_l + \partial_k u_m \partial_l u_m). \tag{10.18}$$

are neglected except those due to the deflection w in direction x_3. The in-plane displacements u_1, u_2 are assumed to be small. The membrane forces N_{kl} are then obtained by an integration across the thickness h

$$\begin{aligned} N_{kl} &= \int_{-h/2}^{h/2} \sigma_{kl} \, \mathrm{d}x_3 = N_{kl}^{\mathrm{L}} + N_{kl}^{\mathrm{N}}; \\ \text{with} \quad N_{kl}^{\mathrm{L}} &= C_{klmn}\varepsilon_{mn}^{\mathrm{L}}, \qquad N_{kl}^{\mathrm{N}} = C_{klmn}\varepsilon_{mn}^{\mathrm{N}}. \end{aligned} \tag{10.19}$$

h is the constant thickness of the plate and C_{klmn} is the elasticity tensor. The bending and twisting moments are as in the linear theory, compare (7.2),

$$m_{kl} = -K_{klmn}\partial_{mn} w. \tag{10.20}$$

whereas there is an additional non-linear term for q_k :

$$\begin{aligned} q_k &= q_k^{\mathrm{L}} + q_k^{\mathrm{N}}; \\ \text{with} \quad q_k^{\mathrm{L}} &= \partial_l m_{kl}, \qquad q_k^{\mathrm{N}} = N_{kl}\partial_l w. \end{aligned} \tag{10.21}$$

The equilibrium equation $\partial_k q_k + f = 0$ is expressed in w

$$D\Delta\Delta w - \partial_k\{N_{kl}\partial_l w\} = f. \tag{10.22}$$

with the isotropic plate rigidity $D = Eh^2/[12(1-\bar{\nu}^2)]$ as in the linear case. f is the transversal load per unit area. The in-plane equilibrium is

$$\partial_l N_{kl} = 0. \tag{10.23}$$

We get finally after introducing a scalar Airy type stress function F with

$$N_{11} = h\partial_{22}F, \qquad N_{12} = N_{21} = -h\partial_{12}F, \qquad N_{22} = h\partial_{11}F \tag{10.24}$$

the *von Kármán equations* for the large deflection of isotropic thin plates

$$\begin{aligned} \Delta\Delta F &= E\left(\partial_{12}w\partial_{12}w - \partial_{11}w\partial_{22}w\right) \\ \Delta\Delta w &= f/D + h/D\left(\partial_{22}F\partial_{11}w - 2\partial_{12}F\partial_{12}w + \partial_{11}F\partial_{22}w\right). \end{aligned} \tag{10.25}$$

The two left-hand sides of these equations are linear and lead to the BIE presented in Chapter 7. The boundary operators and the fundamental solutions for both equations can be taken from the BEM for linear thin plates (Kirchhoff).The non-linear terms are shifted to the right-hand side and can be treated similar to volume forces. They lead to volume integrals for each increment of the iterative solution.

The domain terms in the original space

According to Wang et al. [WJT00], the problem stated above can be solved by a dual reciprocity BEM. They propose to transfer the volume integrals of the right-hand side, which are linked to the non-linearity and to the volume forces, to boundary integrals. The starting point is the following series expansion in global domain interpolation functions ϕ_Ω^i

$$\begin{aligned} \mathcal{L}_1(w,w) &\approx \sum_{i=1}^{N+L} \mathfrak{a}^i \phi_\Omega^i(x); \\ f/D + \mathcal{L}_2(w,F) &\approx \sum_{i=1}^{N+L} \mathfrak{b}^i \phi_\Omega^i(x), \end{aligned} \tag{10.26}$$

with the operators

$$\begin{aligned} \mathcal{L}_1(w,w) &= E\left(\partial_{12}w\partial_{12}w - \partial_{11}w\partial_{22}w\right); \\ \mathcal{L}_2(w,F) &= h/D\left(\partial_{22}F\partial_{11}w - 2\partial_{12}F\partial_{12}w + \partial_{11}F\partial_{22}w\right). \end{aligned}$$

N, L are the boundary and domain collocation points. Crucial for the dual reciprocity approach is the knowledge of ψ_Ω^i with

$$\Delta\Delta\psi_\Omega^i = \phi_\Omega^i. \tag{10.27}$$

Wang et al. [WJT00] propose the following radial basis functions

$$\phi_\Omega^0 = c_1 + c_2|x| + c_3|x|^2 + c_4|x|^3 \tag{10.28}$$

which are translated to each collocation point. A second reciprocity transforms the domain integrals to boundary integrals. Thus, we get for the volume term of the displacement BIE

$$\frac{1}{D}[f + \mathcal{L}_2(w,F)] * W = \frac{1}{D}\sum_{i=1}^{N+L} \mathfrak{b}^i \Delta\Delta\psi_\Omega^i(y) * W \tag{10.29}$$

the following expressions for each collocation point at $x = x^i$, compare (7.15),

$$\begin{aligned}\Delta\Delta\psi_\Omega^i * W &= \kappa^i(x)\psi_\Omega^i(x) \qquad (10.30)\\ &- \int_\Gamma \mathcal{A}_\mathrm{q}\psi_\Omega^i(y)W(x-y)\,\mathrm{d}\Gamma_y - \int_\Gamma \mathcal{A}_\mathrm{m}\psi_\Omega^i(y)\mathcal{A}_\varphi W(x-y)\,\mathrm{d}\Gamma_y\\ &+ \int_\Gamma \mathcal{A}_\varphi\psi_\Omega^i(y)\mathcal{A}_\mathrm{m}W(x-y)\,\mathrm{d}\Gamma_y - \int_\Gamma \psi_\Omega^i(y)\mathcal{A}_\mathrm{q}W(x-y)\,\mathrm{d}\Gamma_y\\ &+ \sum_k \mathcal{A}_\mathrm{c}\psi_\Omega^i(y)W(x-y^k) - \sum_k \psi_\Omega^i(y)\mathcal{A}_\mathrm{c}W(x-y^k).\end{aligned}$$

The solution is then computed iteratively together with the linear BEM. For the details see [WJT00]. The main problems are :

- how to find the ϕ_Ω^i for which ψ_Ω^i in (10.27) are known,
- how to compute the derivatives of ψ_Ω^i at the boundary panels,
- and how to choose the collocation points in the interior domain.

Wang et al. [WJT00] proposed for the second point a second series approximation. To the authors knowledge, there is no Galerkin dual reciprocity method published yet which might simplify the questions mentioned above. Wavelets might be a good choice for ϕ_Ω^i, but this is beyond the scope of this thesis.

10.2.2 Dual reciprocity methods in Fourier space

The evaluation of interpolation functions in Fourier space

In contrast to the traditional approach, the first problem, the evaluation of ψ_Ω^i in (10.27), is very simple in Fourier space. We have for arbitrary global interpolation functions ϕ_Ω^i

$$(\hat{x}_1^2 + \hat{x}_2^2)^2\hat{\psi}_\Omega^i = \hat{\phi}_\Omega^i \qquad \Rightarrow \qquad \hat{\psi}_\Omega^i = \frac{\hat{\phi}_\Omega^i}{(\hat{x}_1^2 + \hat{x}_2^2)^2}. \qquad (10.31)$$

The second problem, the computation of ψ_Ω^i (or its derivatives) at the boundary in (10.30) is also simple if the interpolation functions ϕ_Ω^i are polynomials, trigonometric or exponential functions because the transformations of these functions lead to derivatives of Dirac-distributions. If $\chi^0(x)$ denotes the cutoff distribution of a boundary panel defined in (3.17) or (3.18), and

$\phi_\Omega^i = \sum_k \sum_l c_{kl} x_k^l$ is an arbitrary n-dimensional polynomial, compare (4.17), the values of ϕ_Ω^i at the boundary panel can be computed in Fourier space by

$$\phi_\Omega^i \chi^0 \quad \overset{\mathcal{F}}{\leftrightarrow} \quad \frac{1}{(2\pi)^n} \hat{\phi}_\Omega^i * \hat{\chi}^0 = \sum_k \sum_l c_{kl}\, i^l \partial_k^l \hat{\chi}^0 .$$

Transferring this idea to the dual reciprocity functions ψ_Ω^i and their derivatives $\mathcal{P}^i(\partial)\psi_\Omega^i$ leads to the evaluation of the following convolution

$$\frac{1}{(2\pi)^n} \hat{\mathcal{P}}^i(\hat{x}) \hat{\psi}_\Omega^i * \hat{\chi}^0 = \sum_k \sum_l i^l c_{kl} \int_{\mathbb{R}^2} \frac{\hat{\mathcal{P}}^i(\hat{x}) \delta^{(l)}(\hat{y}_k)}{(\hat{y}_1^2 + \hat{y}_2^2)^2} \hat{\chi}^0(\hat{x} - \hat{y})\, \mathrm{d}\hat{y}, \quad (10.32)$$

which is trivial for all suggested interpolation functions and all straight boundary panels due to the Dirac-distributions in the kernels. The third difficulty, the choice of collocation points, might be less severe when a Galerkin method is chosen, i.e. by choosing ϕ_Ω^i as additional test functions for the interior.

Chapter 11

Wavelets

Alberto Calderón

11.1 Fundamentals of Wavelet Theory

11.1.1 Data compression by wavelets

Although the first mathematical paper on wavelet transform was published by Calderón[1] it took twenty years till Goupillaud, Grossmann and Morlet applied it in an engineering context[2]

Soon afterwards, wavelets became one of the most vivid research areas in mathematics and computational engineering. They have found their way rapidly into industrial applications because of their rather promising properties as means of a very efficient, hierarchical data representation. They are mainly applied in the fields of image processing and signal analysis, although, in the author's opinion, they might be of advantage wherever a large amount of data is handled. The reader interested in the larger context is referred to the webpages http://www.wavelets.org edited by Wim Sweldens.

As the Fourier transform, the wavelet transform is an integral transform technique which has a continuous and a discrete version. The combination with a multiresolution algorithm leads to a fast wavelet transform which might be used as an alternative to the FFT, e.g. Lieb [Lie97]. Finite element methods were developed to represent local characteristics, spectral methods

[1] A.P. Calderón: Intermediate spaces and interpolation, the complex method. *Studia Math.* **24** (1964), 113–190.

[2] P. Goupillaud, A. Grossmann and J.Morlet: Cycle-Octave and related transforms in seismic signal analysis. *Geoexploration* **23**, Elsevier Science Publ. (1984/85), 85–102.

were used for modelling global properties. The wavelet transform is designed for the optimal combination of local and global information.

In 1991, the first paper [BCR91] on wavelets in the context of BEM appeared. The subsequent publications, e.g. [DPS93, BV94, Rat96a, Rat96b, Rat97, LS97, PSS97, PS97, Wan97b, Sch98], concentrating on wavelet discretizations for boundary element methods, succeeded in reducing the work and storage of the discretization scheme from $\mathcal{O}(N^2)$ to $\mathcal{O}(N(\ln N)^a)$ (with a either zero or a small positive integer). The BEM matrices, which are fully populated, are approximated by sparse matrices with a "finger structure" (Fig.11.1). Because of the oscillating character of the wavelets (vanishing moments), some non-essential off-diagonal terms can be neglected on each discretization level of the hierarchical data representation without affecting the convergence rate.

Figure 11.1: The "finger structure" of the approximated BEM-matrix obtained by a wavelet discretization [Sch98]

Unfortunately, this compression algorithm requires a nested sequence of meshes on the boundary Γ and therefore a sufficiently regular geometry of the problem. This is not realistic for practical problems which have in general highly complex geometries. To circumvent this problem, Schmidlin and Schwab [SS00] have proposed a wavelet Galerkin BEM on unstructured meshes which is based on an agglomerated wavelet basis. The algorithm structures the data in clusters gathering data from parts of the geometry which are not necessarily connected. No hierarchical mesh is required.

The method presented in [SS00] is based on the *lifting scheme*, a "second generation wavelet algorithm" developed by Wim Sweldens and uses ideas from other methods for the reduction of complexity in BEM algortihms, i.e. from multipole expansion [Rok85] and panel clustering [HN89].

In this thesis, it is not possible to present a wavelet based BEM in full detail. The intention of the following sections is to give a short introduction to wavelet theory and to show how the compression of BEM matrices can also be obtained for the Fourier BEM.

11.1.2 The wavelet transform

The windowed Fourier transform

The representation of a function f in the original space gives information about its local character. The corresponding Fourier image $\hat{f}$ reflects its global (spectral) properties. The first attempt to combine these two representations has let to the *Gabor transform* (or windowed Fourier transform)[3]

$$\mathcal{G}_\alpha\{f\}(b,\hat{x}) := \int_{-\infty}^{+\infty} f(x) g_\alpha(x-b)\, \mathrm{e}^{-ix\hat{x}}\, \mathrm{d}x, \tag{11.1}$$

where g_α is the window function often chosen as (with a fixed $\alpha > 0$)

$$g_\alpha(x) = \frac{1}{2\sqrt{\pi\alpha}} \exp\left(-\frac{x^2}{4\alpha}\right). \tag{11.2}$$

$\mathcal{G}_\alpha\{f\}$ gives the local and the global information of f. The standard deviation

$$s_\alpha = \frac{1}{\|g_\alpha\|_2} \left[\int_{-\infty}^{+\infty} x^2 g_\alpha^2(x)\, \mathrm{d}x\right]^{1/2} = \sqrt{\alpha}. \tag{11.3}$$

is the measure for the width $L(g_\alpha) = 2\sqrt{\alpha}$ of the x-$\hat{x}$ window. The Fourier transform of g_α, also a Gaussian function, generates the spectral window. The height of the x-$\hat{x}$ window is $L(\hat{g}_\alpha) = 1/(4\sqrt{\alpha})$. Hence, the area of the window has the following constant value (Fig.11.2) :

$$L(g_\alpha) L(\hat{g}_\alpha) = \frac{1}{2}. \tag{11.4}$$

Heisenberg's Uncertainty Principle for an arbitrary window function g

$$L(g)\, L(\hat{g}) \geq \frac{1}{2} \tag{11.5}$$

states that (11.4) is the minimal window area. Therefore, the choice of g_α in (11.2) is optimal. The Gabor transform analyses f by windows with constant

[3]D. Gabor: Theory of Communication, J. IEE (London) **93** (1946), 429–457.

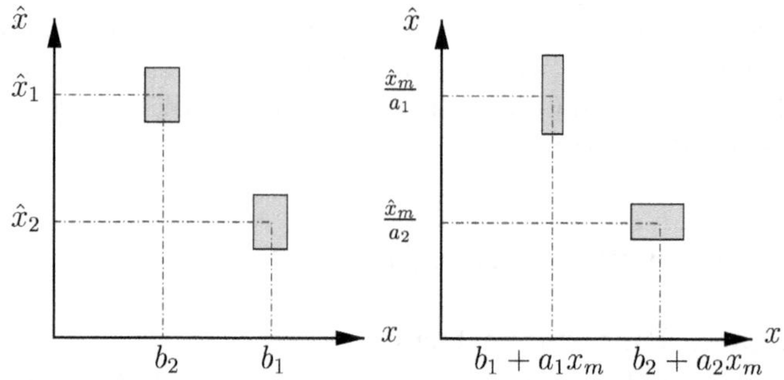

Figure 11.2: The x-$\hat{x}$ window for the Gabor transform (left) and for the wavelet transform (right) [Chu92]

area which is its main disadvantage. High and low oscillating functions cannot be represented adequately. An additional dilational (or scale) parameter can be introduced if $g = \psi$ satisfies [Chu92]

$$\int_{-\infty}^{+\infty} \psi(x)\,\mathrm{d}x = 0 \qquad \overset{\mathcal{F}}{\leftrightarrow} \qquad \hat{\psi}(0) = 0, \tag{11.6}$$

which leads to the continuous wavelet transform.

The continuous wavelet transform

The continuous wavelet transform as a transform which adapts its window size according to the local and global information density of f is defined by

$$\mathcal{W}_{\psi}\{f\}(a,b) := |a|^{-\frac{1}{2}} \int_{-\infty}^{+\infty} f(x)\overline{\psi\left(\frac{x-b}{a}\right)}\,\mathrm{d}x, \qquad a, b \in \mathbb{R}. \tag{11.7}$$

$(\bar{\cdot})$ denotes the complex conjugate. ψ is called *mother wavelet* from which the wavelets $\psi_{a,b}$ are constructed by translation and dilation

$$\psi_{a,b} = |a|^{-\frac{1}{2}}\psi\left(\frac{x-b}{a}\right). \tag{11.8}$$

ψ has to satisfy the admissibility condition [Chu92]

$$C_{\psi} := \int_{-\infty}^{+\infty} \frac{|\hat{\psi}(\hat{x})|^2}{|\hat{x}|}\,\mathrm{d}\hat{x} < \infty. \tag{11.9}$$

The x-$\hat{x}$ window is given by

$$[b + ax_m - ax_r, b + ax_m + ax_r] \times \left[\frac{\hat{x}_m - \hat{x}_r}{a}, \frac{\hat{x}_m + \hat{x}_r}{a}\right], \tag{11.10}$$

where $x_m, \hat{x}_m$ denote the center of the window

$$x_m := \frac{1}{||\psi||_2^2}\int_{-\infty}^{+\infty} x|\psi(x)|^2 \, dx, \qquad \hat{x}_m := \frac{1}{||\hat{\psi}||_2^2}\int_{-\infty}^{+\infty} \hat{x}|\hat{\psi}(\hat{x})|^2 \, d\hat{x},$$

and $x_r, \hat{x}_r$ the corresponding radii ($||.||_2$ is the L^2-Norm) :

$$\begin{aligned} x_r &:= \frac{1}{||\psi||_2}\left\{\int_{-\infty}^{+\infty} (x - x_m)^2|\psi(x)|^2 \, dx\right\}^{1/2}, \\ \hat{x}_r &:= \frac{1}{||\hat{\psi}||_2}\left\{\int_{-\infty}^{+\infty} (\hat{x} - \hat{x}_m)^2|\hat{\psi}(\hat{x})|^2 \, d\hat{x}\right\}^{1/2}. \end{aligned}$$

The reconstruction formula for the wavelet transform is

$$f(x) = \frac{1}{C_\psi}\int_{\mathbb{R}^2} \frac{1}{a^2}\mathcal{W}_\psi\{f\}(a,b)\tilde{\psi}_{a,b} \, da \, db, \tag{11.11}$$

with $\tilde{\psi}_{a,b}$ as the dual wavelet.

The discrete wavelet transform

The discrete wavelet transform with a scale factor $\sigma = 2$ is

$$\begin{aligned} \mathcal{W}_\psi\{f\}(a_j, b_{k,j}) &:= \langle f, \overline{\psi_{k,j}} \rangle = \int_{-\infty}^{+\infty} f(x)\overline{\psi_{k,j}(x)} \, dx; \\ \psi_{k,j}(x) &= 2^{j/2}\psi(2^j x - kb_0), \end{aligned} \tag{11.12}$$

with discrete values for a, b

$$a_j = \sigma^{-j} = 2^{-j}, \qquad b_{k,j} = \frac{k}{2^j}b_0, \qquad j,k \in \mathbb{Z}. \tag{11.13}$$

$b_0 > 0$ is the sampling rate which is often chosen as $b_0 = 1$. For the reconstruction, the wavelet ψ has to satisfy the stability condition, i.e. ψ has to generate a *frame*[4], cf. [Chu92],

$$A||f||_2^2 \leq \sum_{k,j \in \mathbb{Z}} |\langle f, \overline{\psi_{k,j}} \rangle|^2 \leq B||f||_2^2 \tag{11.14}$$

[4] A frame is a collection of vectors a_i of a Hilbert space X so that no element of X is orthogonal to all a_i [Bla98]. In contrast to bases it is not required that the a_i be linearly independent. Hence, frames are redundant, bases not.

for some positive constants $0 < A < B < \infty$ (A, B are called frame bounds). The reconstruction formula is then

$$f(x) = \sum_{k,j\in\mathbb{Z}} c_{k,j}\tilde{\psi}_{k,j}, \qquad c_{k,j} = \langle f, \overline{\psi_{k,j}} \rangle . \tag{11.15}$$

$c_{k,j}$ are the coefficients of the wavelet expansion. In the case of orthogonal wavelets, the dual wavelet $\tilde{\psi}_{k,j}$ is linked to $\psi_{k,j}$ by the biorthogonality relation (δ_{ij} is the Kronecker symbol)

$$\left\langle \psi_{k,j}, \overline{\tilde{\psi}_{k',j'}} \right\rangle = \delta_{k,k'}\delta_{j,j'}, \qquad k, j, l, m \in \mathbb{Z}. \tag{11.16}$$

The dual is obtained in Fourier space by [Chu92]

$$\hat{\tilde{\psi}}(\hat{x}) := \frac{\hat{\psi}(\hat{x})}{\sum_{j_1=-\infty}^{\infty} |\hat{\psi}(\hat{x} + 2\pi j_1)|^2}. \tag{11.17}$$

Alternatively to (11.15), f can be reconstructed by the dual wavelets

$$f(x) = \sum_{k,j\in\mathbb{Z}} d_{k,j}\psi_{k,j}, \qquad d_{k,j} = \left\langle f, \overline{\tilde{\psi}_{k,j}} \right\rangle . \tag{11.18}$$

11.1.3 Multiresolution analysis

Decomposition into a sequence of hierarchical structured details

Around 1986, the *multiresolution analysis* (MRA), a new method for performing discrete wavelet analysis, was developed [Mal89, Mey90]. The coefficients $c_{j,k}$ (or $d_{j,k}$) on the finest level (smallest value a_j) are computed explicitely. From these, all coefficients on all coarser levels can be constructed by a recursive algorithm. Thus, the total computational effort is reduced remarkably; the wavelet analysis based on a multiresolution algorithm is also called *fast wavelet transform* (FWT).

A function f_0 is sampled on the finest level $j = 0$ at regular intervals $\Delta x = \xi$. Then this sequence is split into a coarser version f_{-1} at the coarser scale with $\Delta x = 2\xi$ and into the complementary details d_{-1} at scale $\Delta x = \xi$. The repetition of this process leads to a sequence $f_0, f_{-1}, f_{-2}, \dots$ of coarser and coarser approximations with complementary sequences $d_0, d_{-1}, d_{-2}, \dots$ of details removed at every scale. Each detail d_j is then described by a superposition of wavelets $\psi_{k,j}, k \in \mathbb{Z}$ with a special mother wavelet determined recursively. The original function f_0 can be recovered after N iterations by summing all details and adding the coarsest approximation f_{-N}

$$f_0 = f_{-N} + d_{-N+1} + d_{-N+2} + \cdots + d_0. \tag{11.19}$$

Nested sequence of function spaces for the approximations f_j

The MRA constructs a nested sequence of vector spaces

$$\{0\} \subset \cdots \subset V_2 \subset V_1 \subset V_0 \subset V_{-1} \subset V_{-2} \subset \cdots \subset L^2(\mathbb{R}) \qquad (11.20)$$

satisfying the following properties

$$\bigcap_{j\in\mathbb{Z}} V_j = \{0\}, \qquad \operatorname{clos} \bigcup_{j\in\mathbb{Z}} V_j = L^2(\mathbb{R}). \qquad (11.21)$$

where clos denotes the closure of a vector space. This means that the spaces $V_j, j \in \mathbb{Z}$ have only the trivial function $\phi = 0$ in common and that all elements of L^2 can be represented by functions of these spaces. The space V_{j-1} is obtained from V_j by scaling

$$\phi(x) \in V_j \quad \Leftrightarrow \quad \phi(2x) \in V_{j-1}, \qquad (11.22)$$

and the spaces V_j are closed with respect to translation

$$\phi(x) \in V_j \quad \Leftrightarrow \quad \phi(x-k) \in V_j, \qquad k \in \mathbb{Z}. \qquad (11.23)$$

The space V_j is spanned by the *scaling functions* $\phi_{k,j}(x), k \in \mathbb{Z}$ which are used to approximate f_j on the j-th level of coarseness[5].

Nested sequence of function spaces for the details d_j

The details d_j on level j are the complement of the coarse approximation f_j. Hence, we define function spaces W_j for the details which are complementary to V_j. They satisfy

$$V_{j-1} = V_j + W_j \qquad (11.24)$$

which implies that V_j can be expressed by the sum of the spaces of details W_l and V_N on the coarsest level

$$V_j = V_N + \sum_l W_l. \qquad (11.25)$$

The spaces W_j are spanned by wavelets $\psi_{k,j}, k \in \mathbb{Z}$, i.e. the details d_j are expressed by a sum of wavelets $\psi_{k,j}, k \in \mathbb{Z}$ while the coarsest approximation f_N of f is described by $\phi_{k,N}$.

[5] The scaling functions $\phi_{k,j}(x), k \in \mathbb{Z}$ are the *Riesz basis* of the vector space V_j [LMR94].

Mother wavelets ψ constructed from mother scaling functions ϕ

The MRA starts by defining a *mother scaling function* ϕ which has to satisfy as a necessary and sufficient condition the scaling relation [Chu92]

$$\phi(x) = \sum_{k=-\infty}^{\infty} \mathfrak{h}_k \phi(2x - k). \tag{11.26}$$

The *scaling functions* $\phi_{k,j} \in V_j$ are obtained by translating and dilating ϕ :

$$\phi_{k,j}(x) = 2^{-j/2}\phi(2^{-j}x - k). \tag{11.27}$$

They are used as a Riesz basis for V_j. The dual scaling functions $\tilde{\phi}_{k,j}$ defined by the biorthogonality relation on level j

$$\left\langle \phi_{k,j}, \overline{\tilde{\phi}_{k',j}} \right\rangle = \delta_{k,k'} \tag{11.28}$$

lead to the dual MRA. The scaling functions are orthonormal at every scale but not from one scale to the other. The Fourier transform of the scaling equation (11.26) is

$$\hat{\phi}(\hat{x}) = \hat{p}(\hat{x}/2)\hat{\phi}(\hat{x}/2), \qquad \text{with} \qquad \hat{p}(\hat{x}) = \frac{1}{2}\sum_{k=-\infty}^{\infty} e^{-ik\hat{x}}\mathfrak{h}_k. \tag{11.29}$$

The numerical realizations of a MRA algorithm are based on the coefficients $\mathfrak{h}_k$ from equation (11.26). Therefore, $\hat{p}$ is called *generating function* of the MRA. The mother wavelet is then constructed from the mother scaling function in Fourier space by, see for example Blatter [Bla98] for the justification,

$$\hat{\psi}(\hat{x}) = e^{i\hat{x}/2}\overline{\hat{p}(\hat{x}/2 + \pi)}\,\hat{\phi}(\hat{x}/2). \tag{11.30}$$

The relation in the original space is given by Blatter as

$$\psi(x) = \sum_{k=-\infty}^{\infty} \mathfrak{g}_k \phi(2x - k), \qquad \text{with} \qquad \mathfrak{g}_k = (-1)^k \overline{\mathfrak{h}_{2N-1-k}}. \tag{11.31}$$

The algorithm for the fast wavelet transform

The two fundamental recursive relations for the fast wavelet algorithm

$$\phi_{k,j} = \sum_l \mathfrak{h}_l \phi_{2k-l,j-1}, \qquad \psi_{k,j} = \sum_l \mathfrak{g}_l \phi_{2k-l,j-1} \tag{11.32}$$

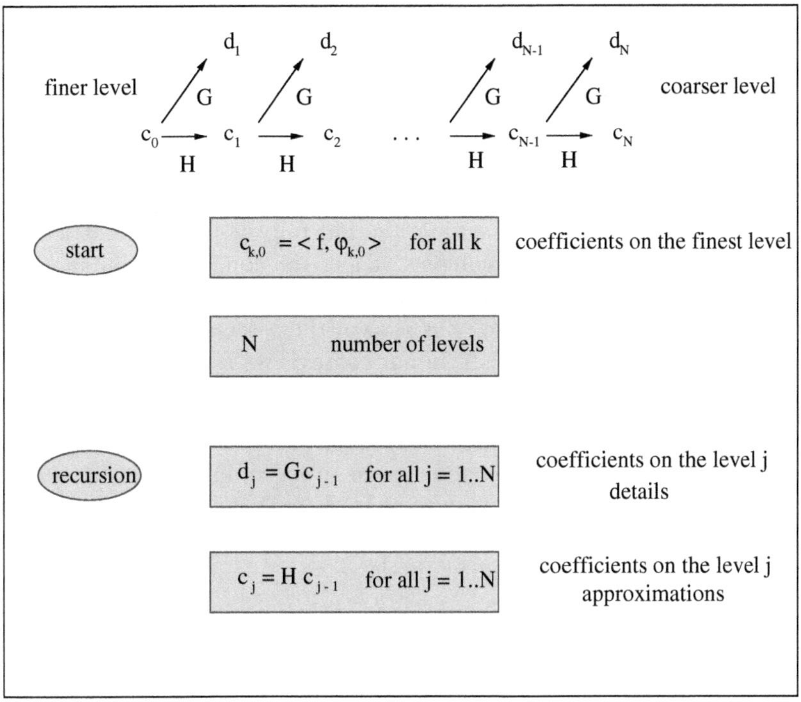

Figure 11.3: The algorithm of the fast wavelet transform

are obtained from (11.26) and (11.31) [Bla98]. The analysis of a function $f \in L^2$ starts at the finest level $j = 0$ where all coefficients

$$c_{k,0} = \left\langle f, \overline{\phi_{k,0}} \right\rangle \tag{11.33}$$

have to be computed explicitely. Then, the coefficients $c_{k,j}, d_{k,j}$ for the coarser levels are evaluated recursively by [LMR94]

$$c_{k,j} = \sum_{l \in \mathbb{Z}} \mathfrak{h}_{l-2k} c_{l,j-1}, \qquad d_{k,j} = \sum_{l \in \mathbb{Z}} \mathfrak{g}_{l-2k} c_{l,j-1}, \tag{11.34}$$

or in operator notation (Fig.11.3)

$$c_j = H c_{j-1}, \qquad d_j = G c_{j-1}, \tag{11.35}$$

with $Hc_{j-1} = \sum_{l \in \mathbb{Z}} \mathfrak{h}_{l-2k} c_{l,j-1}$ and $Gc_{j-1} = \sum_{l \in \mathbb{Z}} \mathfrak{g}_{l-2k} c_{l,j-1}$. The original function can be reconstructed by

$$c_{j-1} = H^* c_j, \qquad d_{j-1} = G^* c_j, \tag{11.36}$$

with the adjoint operators defined by $H^*c_j = \sum_{l\in\mathbb{Z}} \mathfrak{h}_{k-2l}c_l$ and $G^*c_j = \sum_{l\in\mathbb{Z}} \mathfrak{g}_{k-2l}c_l$.

11.2 Wavelet Galerkin Discretization

The Galerkin BEM leads to symmetric, but fully populated matrices which becomes relevant for large problems. Then, the computational effort is not determined by the computation of the matrix entries but by the inversion of the matrix. As recent research has shown, there are several possibilities for compressing the BEM matrices. In our context, we present first ideas on the compression by wavelets for the Fourier BEM which are closely related to the work done by Schneider [Sch98], Schwab [LS97, SS00] and coworkers.

It is far beyond the scope intended here to give a complete presentation of the compression algorithms. We only want to show that the existing approaches can be transferred easily to the Fourier BEM. The explicit realization might be the subject of an independent thesis.

11.2.1 Matrix compression in the original space

Wavelets as trial and test functions

With the exception of a very recent prepublication [SS00], all wavelet based BEMs found in the literature require a sufficiently regular and structured geometry. They assume that the boundary can be decomposed into N_0 curvilinear, either quadrilateral or triangular surface pieces Γ_j. The corresponding reference domain Γ_j is either the unit triangle or the unit square, e.g. [PS97]. The wavelet analysis is then constructed on this reference domain : Γ_0 is divided into a sequence of $4^l, l = 1..L$ subsquares or subtriangles $\{\Gamma_{k,l}\}$ by successively halving the sides l times. On these, θ_d discontinuous piecewise polynomials of degree d are defined.

This ends up in $j = 1..N_0$ surface pieces, $l = 1..L$ levels of subdivisions, $k = 1..4^l$ subelements on each level l, and $m = 1..\theta_d$ scaling functions for each subelement. ϕ_I denotes the scaling function with the multiindex $I = (j,k,l,m)$ and ψ_J is the complementary wavelet. Note that each scaling function leads to 3 wavelets, i.e. we have for the wavelets $m = 1..3\theta_d$ and $J = (j,k,l,3m)$. The wavelets are piecewise polynomials in local coordinates [PS97]. The left part of Fig.11.4 shows an example for a linear wavelet on a triangular panel which is given by [PSS97]. The right part shows the real and imaginary part of the Fourier transformed wavelet.

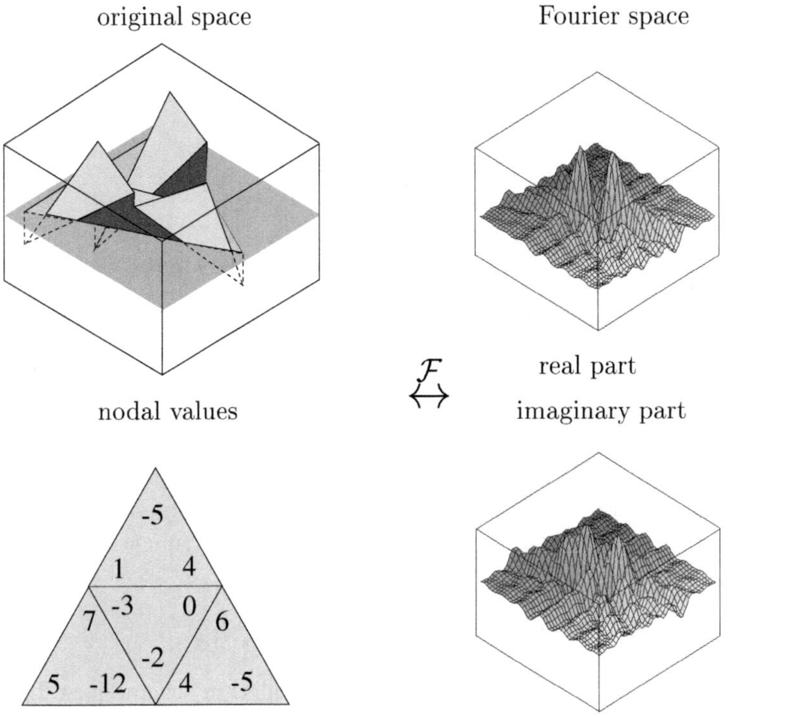

Figure 11.4: A piecewise linear wavelet ψ in the original (left) and in Fourier space (right)

Sketch of the multiscale Galerkin discretization

Most of the approaches already published are based on a Galerkin BIE of the following form

$$\left\langle \phi^j, \mathcal{B}u_\chi \right\rangle = \left\langle \phi^j, f * \mathcal{B}U \right\rangle + \sum_i \left\langle \phi^j, \phi^i * \mathcal{A}U \right\rangle, \tag{11.37}$$

with some differential operators $\mathcal{A}, \mathcal{B}$, compare for example (6.19). The trial and the test functions are then replaced by wavelets $\psi_J, \psi_{J'}$, i.e. the boundary quantities u, t are approximated by

$$u^L \approx \sum_{l=0}^{L} \sum_J \mathrm{u}_J^L \psi_J, \qquad t^L \approx \sum_{l=0}^{L} \sum_J \mathrm{t}_J^L \psi_J. \tag{11.38}$$

The matrix entries are computed from

$$G^L_{J'J} = \langle \psi_{J'}, \psi_J * \mathcal{A}U \rangle , \qquad H^L_{J'J} = \langle \psi_{J'}, \psi_J * U \rangle , \tag{11.39}$$

An estimate for the entries is given by [PS97][6]

$$|G^L_{J'J}| \leq C d_{J'J}^{-2(d+2)} 2^{-(d+2)(l+l')} \tag{11.41}$$

which depends on the distance between two wavelets defined by

$$d_{J'J} = \text{dist}(S_J, S_{J'}), \qquad \text{with} \quad S_J = \{x \in \Gamma; \psi_J(x) \neq 0\}. \tag{11.42}$$

C is a constant independent of the levels l', l. If the two wavelets are seperated by a distance $d_{J'J}$ larger than a certain truncation parameter $\Delta_{l',l}$ (depending on the level l and level l') the entries can be neglected without affecting the asymptotic convergence rate [PS97, Sch98]. Hence, as long as the geometry is sufficiently structured, most of the entries of the stiffness matrix can be a priori replaced with zero. The minimal value for the truncation parameter is determined by [PSS97]

$$\Delta_{l',l} \geq \max\{a 2^{-L} 2^{\alpha(L-l)} 2^{\alpha'(L-l')}, 2^{-l}, 2^{-l'}\}, \tag{11.43}$$

with some number $a \geq 1$. The compressed Galerkin scheme is stable if a is sufficiently large. The values α, α' are given by [PS97]

$$\alpha \geq \frac{s+d+1}{2(d+1)}, \qquad \alpha' \geq \frac{s'+d+1}{2(d+1)},$$

with $s, s' \in [0, d+1)$.

The presentation given here is only a rough sketch, for a realization it is essential to study the literature cited above.

11.2.2 Matrix compression in Fourier space

Transform of the trial and test wavelets

As long as the wavelets are defined on straight elements, the wavelet discretization presented in the last section can be transferred to the Fourier

[6] This is essentially due to the vanishing moment properties of the wavelets [Sch98]

$$\int x^\alpha \psi_J \, \mathrm{d}x = 0, \qquad \text{for } |\alpha| \leq d. \tag{11.40}$$

BEM. The truncation criterion (11.43) and the distances $d_{J',J}$ between the wavelets remain valid. We only have to find the transformed wavelets and the transformations for the scaling functions on the coarsest level. They are defined as piecewise polynomials on triangular or quadrilateral reference elements. Hence, the transformed trial and test wavelets $\hat{\psi}_J, \hat{\psi}_{J'}$ are obtained by a combination of (4.15) to (4.17). The real and imaginary part of the transformed sample wavelet are visualized on the right-hand side of Fig.11.4.

Transform of the multiscale Galerkin BIE

The remaining non-zero matrix entries can be computed by the transform of (11.39)

$$G^L_{J'J} = \frac{1}{(2\pi)^n} \left\langle \hat{\psi}_{J'}(-\hat{x}), \hat{\psi}_J \hat{A} \hat{U} \right\rangle, \qquad H^L_{J'J} = \frac{1}{(2\pi)^n} \left\langle \hat{\psi}_{J'}(-\hat{x}), \hat{\psi}_J \hat{U} \right\rangle. \quad (11.44)$$

The vanishing moment condition (11.6)

$$\hat{\psi}(0) = 0$$

as fundamental condition for wavelets leads to an annihilation of the singularities of the fundamental solution at $\hat{x} = 0$ (in the case of elliptic operators). We have for example for the Laplacian in two dimensions $\hat{U} = -(\hat{x}_1^2 + \hat{x}_2^2)^{-1}$ and hence for the kernel

$$\lim_{|\hat{x}| \to 0} \hat{\psi}_{J'}(-\hat{x}) \hat{\psi}_J \hat{U} = 0. \quad (11.45)$$

It is only at the coarsest level that the singularity remains at $\hat{x} = 0$, because the scalar product with two scaling functions is not vanishing at $\hat{x} = 0$:

$$\lim_{|\hat{x}| \to 0} \hat{\phi}_N(-\hat{x}) \hat{\phi}_N \hat{U} \neq 0. \quad (11.46)$$

These ideas remain to be developed. Here, they are intended as a preview of the future development of the Fourier BEM.

Chapter 12

Conclusions

12.1 Results

This thesis is the first principle work on the generalization of the boundary element method by spatial and temporal Fourier transform. The traditional Galerkin boundary integral equations are reformulated by means of the convolution theorem and Parseval's identity. In contrast to the traditional method, all quantities, the trial and test functions as well as the fundamental solution and its derivatives, are only required in Fourier space. The matrices are evaluated directly, no inverse transform is necessary.

For complicated physical problems (e.g. anisotropic elasticity), the fundamental solution is only known explicitely in Fourier space. Hence, approaches via traditional BEM encounter a lot of difficulties which are avoided in the method presented here. Therefore, the field of applications of the BEM is enormously enlarged.

The new approach is valid for all linear problems where the material parameters do not depend on location or time. The examples presented were taken from stationary heat conduction, elasticity, thin and thick plates with or without Winkler foundation. Transient and stationary dynamic problems were discussed for elasticity, poroelasticity, plates, and thermoelasticity.

Geometrical and physical non-linearities can be modelled by iterative procedures where the non-linear contributions are shifted to the right-hand side of the equations and treated as initial volume terms for each increment.

The Galerkin BEM leads to symmetric, but fully populated matrices. It was shown that recent methods based on the fast wavelet transform to sparsify these matrices can be transferred to Fourier BEM.

For the realization of Fourier BEM, a rigorous distributional representation of the boundary element method was developed. It enables the correct treatment of the jumps and singularities occuring in the boundary integral equations of traditional BEM or Fourier BEM. The distributional concept is necessary for the differentiation of boundary integral equations required for the symmetric Galerkin BEM. It results in demonstrating that all singularities (which are often non-integrable) cancel in the original space if the integrals are combined adequately.

12.2 Open Questions

There are three main open questions: Firstly, all examples were computed by integration routines which are not really optimized for the oscillating kernels occurring in the integrals of the Fourier BEM. First ideas may be taken from a paper published by Averbuch et al. [ABC$^+$00] on the computation of oscillatory integrals by wavelets.

And secondly, the Fourier BEM is still restricted to cases where the material parameters are constant with respect to time and space. The character of the differential equations changes totally if this assumption is abandoned. A way of establishing a BEM for this kind of problems may be read in the work of Pomp [Pom98].

Finally, the approach presented here is only valid for straight elements. For the implementation of curved elements, the approach of Andrä [AS97] might be useful. The curved elements are approximated locally by tangential straight elements of the same proportions. The kernel and the Jacobian are expanded in Taylor series.

Acknowledgements

The author would like to express his gratitude for the grants which have supported this thesis. The grant from the French Alfred-Kastler foundation, which provided support for a stay at the Ecole Polytechnique, Laboratoire de Mécanique des Solides (LMS), in Paris, gave me the opportunity to develop the initial ideas. I am particularly indebted to Prof. Marc Bonnet for his criticism and comments and the benefit of his experience in the field of boundary element methods. They provided the real incentive to my work.

Further encouraged by the fruitful and open discussions and by the kindness of all members of the LMS, I went back to Munich. Prof. Harry Grundmann, who had already supervised my PhD thesis, hosted me during the following two years. Once more I am deeply grateful for his personal and professional support. Better working conditions cannot be found. The financial support, a grant especially for the postdoctoral thesis, from the German Research Foundation (DFG) is gratefully acknowledged.

I am also indebted to many other people working in the field of boundary element methods. First of all to Andreas Pomp who helped me to understand some deeper mathematical items related to my work. Prof. Joachim Gwinner from the Universität der Bundeswehr in Munich was always open to my problems. The constructive criticism of Prof. Friedel Hartmann helped me to complete my work. Prof. Kuhn in Erlangen, Prof. Wendland in Stuttgart, and Prof. Maier in Milan gave me the opportunity to present my work. Without these discussions I would not have progressed as fast as I did. Prof. Guiggiani in Pisa and Attilio Frangi in Milan helped me a lot with their questions and comments. I would like to thank Prof. Ernst Rank for pushing me more towards engineering applications. I was sometimes lost in self-generated artificial problems.

I would like to thank my family, my wife and my children, for their flexibility and patience, and my father for his critical reading.

Appendix A

Glossary

A.1 Distribution theory

support: The support of a function $\phi(x)$

$$\operatorname{supp}\{\phi\} = \overline{\{x \in \mathbb{R}^n; \phi(x) \neq 0\}}$$

is defined as the closure of the set of all points $x \in \mathbb{R}^n$ where ϕ is not vanishing. Single points x with $\phi(x) = 0$ are included in this set.

C^m-continuous: A function ϕ is called C^m-continuous if it is continuous up to its m-th derivative.

differentiation: $\partial_k \phi$ denotes the k-th differentiation of a function ϕ with respect to x_k. ∂^α with the multiindex α standing for

$$\partial^\alpha = \partial_1^{\alpha_1} \partial_2^{\alpha_2} \dots \partial_n^{\alpha_n}, \qquad |\alpha| = \sum_{k=1}^{n} \alpha_k, \qquad \alpha_k \in \mathbb{N}_0 .$$

Hölder-continuous: A function $\phi \in C^{m,\lambda}$ is Hölder continuous of order λ if it satisfies for two points $x = \xi_1, \xi_2$ up to its $|\alpha|$-th derivative ($|\alpha| \leq m$) the condition (C is an arbitrary constant)

$$|\partial^\alpha \phi(\xi_1) - \partial^\alpha \phi(\xi_2)| \leq C |\xi_1 - \xi_2|^\lambda, \qquad 0 < \lambda \leq 1.$$

Lipschitz-continuous: A function ϕ is called Lipschitz continuous if it is Hölder-continuous with $\lambda = 1$.

C^1 boundary: A boundary is a C^1 boundary if for every boundary point $x_0 \in \partial\Omega$ one can find a C^1 function ψ in a neighbourhood of x_0 with

$$\psi(x_0) = 0, \qquad d\psi(x_0) \neq 0.$$

L^p-spaces: The Banach spaces $L^p, p \geq 1$ are defined as the space of functions ϕ which are to the p-th power integrable. They are endowed with the norm

$$||\phi||_p = \left[\int_{\mathbb{R}^n} |\phi(x)|^p \, dx\right]^{1/p} < \infty.$$

For $p = 2$, the scalar product of two functions $\phi_1\phi_2$ can be defined

$$\langle\phi_1, \phi_2\rangle = \int_{\mathbb{R}^n} \phi_1(x)\bar{\phi}_2(x) \, dx.$$

The L^2 is a Hilbert space[1].

Sobolev spaces: Sobolev spaces $W^{m,p}$ are spaces of functions ϕ which are up to their m-th derivative in L^p

$$W^{m,p} = \{\phi; \partial^\alpha \in L^p(\Omega); \text{for all } |\alpha| \leq m\}.$$

If $p = 2$ the spaces $W^{m,2}$ are notated by H^m.

trace operator: For a $\mathcal{C}^m$-regular domain, the trace operator γ^s of order $s \leq m$ yields the boundary values of a function ϕ up to the k-th derivative (k as an integer smaller than $s - 1/2$). We have

$$\gamma^s : \phi \to \gamma^s\phi = (\gamma_0\phi, \dots \gamma_k\phi),$$

with the traces as the normal derivatives

$$\gamma_j\phi = \left.\frac{\partial^j\phi}{\partial\nu^j}\right|_{\partial\Omega}, \qquad \text{for all } j = 0, \dots, k.$$

generalized functions: The set of generalized functions u include all linear and continuous functionals. They are defined via some test functions ϕ

$$u(\phi) = \langle u, \phi\rangle < \infty.$$

Hence the properties of the test functions define the set of generalized functions. In the following, the four main types are given with the corresponding sets of test functions.

[1]In the context of the present thesis, the scalar product is defined without the conjugate complex. The functions are real-valued.

distributions[2]: The test functions $\phi(x) \in \mathcal{D}(\Omega) = C_0^\infty(\Omega)$ are bounded, possess a compact support, and are infinitely often continuously differentiable. They and all their derivatives vanish at the boundary. To demonstrate that this set is not empty, Hörmander gives the following example : For all closed balls with the center $x_0 \in \Omega \subset \mathbb{R}^n$ and the radius $r > 0$ which are totally in the domain Ω, there exists an exponential test function (Fig.A.1 with $x_0 = 0, r = 1$).

$$\phi(x) = \begin{cases} \mathrm{e}^{-\frac{r^2}{r^2-|x-x_0|^2}} & \ldots \quad |x| < r \\ 0 & \ldots \quad |x| > r. \end{cases}$$

A distribution $u \in \mathcal{D}'(\Omega)$ is defined by the scalar product with the test function ϕ

$$u(\phi) = \langle u, \phi \rangle = \int_{\mathbb{R}^n} u(x)\phi(x)\,\mathrm{d}x < \infty.$$

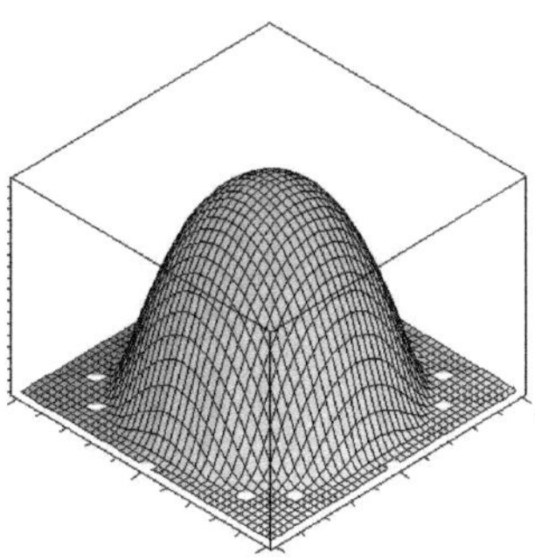

Figure A.1: *Two-dimensional test function* $\phi(x) \in C_0^\infty(\Omega)$

tempered distribution: The set of test functions can be enlarged to all functions $\phi \in \mathcal{S}(\mathbb{R}^n)$ which decline exponentially including all their derivatives (the Gaussian is an example). The corresponding (dual) distributions $u \in \mathcal{S}'(\mathbb{R}^n)$ are called tempered distributions. Because we have weakened the conditions on the test functions, the set of tempered distributions is smaller than that of distributions. The set $\mathcal{S}'(\mathbb{R}^n)$ is closed with respect to the Fourier transform which was the historical reason for their definition.

distribution with compact support: If the condition at infinity is abandoned one can define test functions $\phi \in C^\infty(\mathbb{R}^n) = \mathcal{E}(\mathbb{R}^n)$. To avoid difficulties at infinity, the corresponding distributions $u \in \mathcal{E}'(\mathbb{R}^n)$ have to have compact support. Their Fourier transform is a C^∞-function without any local singularities.

ultradistributions: The Fourier transform of a distribution leads not evidently to a distribution. For example, the Fourier transform of $e^{x_1}, x_1 \in \mathbb{R}$ yields $\delta(\hat{x}_1 - i), i = \sqrt{-1}$ which is not a distribution. Because the Fourier transform of generalized functions is defined via the Fourier transform of the test functions, we get the test functions $\phi \in \mathcal{Z}(\mathbb{C}^n)$ for the ultradistributions $u \in \mathcal{Z}'(\mathbb{C}^n)$ by Fourier transform of the test functions $\phi \in \mathcal{D}(\Omega)$ of the distributions.

generalized differentiation: The differentiation of generalized functions is defined via partial integration, i.e. the differential operator is shifted to the test functions.

$$\langle \partial_k u, \phi \rangle = -\langle u, \partial_k \phi \rangle, \qquad \phi \in C_0^\infty.$$

Because of the definition of the test functions ϕ, the distributions are infinitely often differentiable. Jumps and singularities can be differentiated.

differentiation of a product: If the multiplication of two distributions u_1, u_2 is defined, we have for the differentiation of a product (Leibniz)

$$\partial_k(u_1 u_2) = u_2 \partial_k u_1 + u_1 \partial_k u_2.$$

convolution of distributions: The convolution

$$u_1 * u_2 := \int_{\mathbb{R}^n} u_1(x-y) u_2(y)\, \mathrm{d}y; \qquad x, y \in \mathbb{R}^n.$$

of two distributions u_1 and u_2 is defined as long as either u_1 or u_2 has compact support [Hör90]. We have for $u_1 \in \mathcal{D}'(\Omega)$ and $u_2, u_3 \in \mathcal{E}'(\mathbb{R}^n)$

$$\begin{array}{lrcl} \text{commutation} & u_1 * u_2 & = & u_2 * u_1 \\ \text{association} & u_1 * (u_2 * u_3) & = & (u_1 * u_2) * u_3 \end{array}$$

differentiation of a convolution product: We have for the differentiation of a convolutional product

$$\partial^\alpha(u_1 * u_2) = (\partial^\alpha u_1) * u_2 = u_1 * (\partial^\alpha u_2),$$

which means that we can exchange differentiation and the integration related to convolution [Hör90].

Cauchy principal value: Some singular integrals can be regularized by taking into account the fact that the singularities cancel. If for example, the integral is singular at $x_0 \in [a, b]$ then the principal value is obtained from

$$\text{v.p.} \int_a^b \phi(x)\,\mathrm{d}x = \lim_{\varepsilon \to 0^+} \left\{ \int_a^{x_0-\varepsilon} \phi(x)\,\mathrm{d}x + \int_{x_0+\varepsilon}^b \phi(x)\,\mathrm{d}x \right\}.$$

The two integrals are infinite, but the limit of the sum is finite.

Finite-part integral: The finite part (Hadamards principal value) as an extension of the concept of Cauchy's principal value is defined as the regular part of a singular integral. If an integration leads to a function

$$g(\varepsilon) = \lim_{\varepsilon \to 0} \sum_{j=1}^{N} \frac{b_j}{\varepsilon^{a_j}} + b_{N+1} \ln \varepsilon + b_{N+2} + \mathcal{O}(1)$$

then the term b_{n+2} is called the finite part of $g(\varepsilon)$, [McL00],

$$\text{p.f. } g(\varepsilon) = b_{N+2}.$$

This means that all singular contributions are neglected.

Gauss-Green formula: If a domain Ω is described by a cutoff distribution χ we have for the product of χ with a distribution u, [Hör90]

$$\partial_j(u\chi) = (\partial u)\chi - u\nu_j\,\mathrm{d}\Gamma.$$

$\mathrm{d}\Gamma$ is the Euclidian surface measure on $\partial\Omega$ and ν the exterior unit normal. The integration weighted with $\phi = 1$ of this equality yields the theorem of Gauss

$$\int_\Omega 1 \cdot \partial_j(u\chi)\,\mathrm{d}\Omega = -\int_\Omega (u\chi)\partial_j\{1\}\,\mathrm{d}\Omega = 0 = \int_\Omega \partial_j u\,\mathrm{d}\Omega - \int_\Gamma u\nu_j\,\mathrm{d}\Gamma.$$

Fourier transform of tempered distributions: For tempered distributions $u \in \mathcal{S}'$, the Fourier transform $\hat{u}$ is defined by the transform of the test function ϕ

$$\hat{u}(\phi) = u(\hat{\phi}), \qquad \phi \in \mathcal{S}.$$

Parseval's identity: The invariance of the scalar product concerning the Fourier transform is called Parseval's idenitity

$$\begin{aligned} \langle u_1, u_2 \rangle &= \frac{1}{(2\pi)^n} \langle \hat{u}_1, \hat{u}_2 \rangle ; \\ \int_{\mathbb{R}^n} u_1(x)\bar{u}_2(x)\,\mathrm{d}x &= \frac{1}{(2\pi)^n} \int_{\mathbb{R}^n} \hat{u}_1(\hat{x})\bar{\hat{u}}_2(\hat{x})\,\mathrm{d}\hat{x}. \end{aligned}$$

Alternatively for real valued distributions, we have

$$\int_{\mathbb{R}^n} u_1(x)\bar{u}_2(x)\,\mathrm{d}x \;=\; \frac{1}{(2\pi)^n}\int_{\mathbb{R}^n} \hat{u}_1(-\hat{x})\hat{u}_2(\hat{x})\,\mathrm{d}\hat{x}.$$

Fourier transform of distributions: The transform $\hat{u} \in \mathcal{Z}'(\mathbb{C}^n)$ of a distribution $u \in \mathcal{D}'(\Omega)$ is defined via Parseval's identity

$$\langle u, \phi_{\mathcal{D}}\rangle = \frac{1}{(2\pi)^n}\left\langle \hat{u}, \hat{\phi}_{\mathcal{Z}}\right\rangle,$$

This means that it is constructed from the Fourier transform $\hat{u}_{\mathcal{Z}} \in \mathcal{Z}(\mathbb{C}^n)$ of the test function $\phi_{\mathcal{D}} \in \mathcal{D}(\Omega)$. Hence we have also defined the Fourier transform of ultradistributions.

theorem of Paley-Wiener: If the support of a distribution u is contained in a convex and compact domain $\Omega \Subset \mathbb{R}^n$ with a cutoff distribution χ then the transform of u has no local singularities [Hör90]. Consequently, every distribution with compact support has a Fourier transform which is an entire analytic function in $\mathbb{C}^n$.

jumps and singularities: A distribution has singularities of degree $s \in \mathbb{R}$ if its transform behaves at infinity like $\mathcal{O}(|\hat{x}|^s)$. The Dirac-distribution is of degree $s = 0$ and jumps are singularities of degree $s = -1$.

A.2 Boundary Element Method

boundary integral equations: The boundary integral equations are obtained by a Somigliana identity where the source points are put on the boundary itself.

direct and indirect BEM: In the direct BEM, the real physical boundary quantities are used. The indirect BEM defines fictitious sources of unknown density on the boundary.

fundamental solution: The fundamental solution U is the response of the infinite medium to a single source $f = \delta$

$$\mathcal{P}(\partial)U = \delta.$$

It is the inverse of the differential operator $\mathcal{P}(\partial)$.

Green's function: A solution U of $\mathcal{P}(\partial)U = \delta$ is called Green's function if its takes into account particular boundary conditions.

reciprocity relation: If the virtual work is computed for a first state of equilibrium by choosing a second state as virtual state, and if a second virtual work is formulated by exchanging the two states, the two virtual works will be equal. This equality is called reciprocity relation and is also known as Betti's theorem.

Somigliana's identity: If one state in a reciprocity relation is chosen as the state related to the fundamental solution, the resulting equality is called Somigliana's identity.

trial and test functions: The known and unknown quantities are approximated by trial functions which are located on the boundary. The equations are then weighted (or tested) by test functions which are either identical to the trial functions (Galerkin approach) or which are Dirac-distributions (collocation approach). In engineering mechanics, the test functions are often called virtual functions.

single layer potential: The single layer potential is defined as the convolution of a boundary quantity t with the fundamental solution U

$$V\{t\} = \int_\Gamma t(y)U(x-y)\,\mathrm{d}y.$$

V is the operator of the single layer potential.

double layer potential: The double layer operator K yields a convolution with the normal derivative (with respect to y) of the fundamental solution

$$K\{u\} = \int_\Gamma u(y)\frac{\partial}{\partial \nu_y}U(x-y)\,\mathrm{d}y.$$

Its adjoint operator is obtained by taking the normal derivative at the point x

$$K^*\{t\} = \int_\Gamma t(y)\frac{\partial}{\partial \nu_x}U(x-y)\,\mathrm{d}y.$$

hypersingular potential: The hypersingular operator W results from a double differentiation

$$W\{u\} = -\frac{\partial}{\partial \nu_x}\int_\Gamma u(y)\frac{\partial}{\partial \nu_y}U(x-y)\,\mathrm{d}y.$$

Due to its singularity, this integral cannot be differentiated directly under the integral sign. Taking into account the differentiation of a convolution product (see above) we have

$$W\{u\} = u(x) * [-\nu_x \cdot \nabla(\nu_y \cdot \nabla U(x))].$$

where the support of u is restricted to the boundary.

free term: Because of the sifting property of the Dirac-distribution, the Somigliana identity reduces one volume integral to a single value which is called free term. For the direct BEM, this single term takes into account the shape and the regularity of the boundary.

source and field point: In the case of the collocation BEM, the boundary integral equations are solved for discrete source points of the boundary by setting the Dirac-distribution (the "source" for the fundamental solution) on these points. The integration is computed with respect to the field (observation) point.

singularities (2D): We distinguish the following types of singularities:

	weak singular	strong singular	hypersingular
$\mathbb{R}^2$	$\mathcal{O}(\ln \lvert x\rvert)$	$\mathcal{O}(\lvert x\rvert^{-1})$	$\mathcal{O}(\lvert x\rvert^{-2})$
$\mathbb{R}^3$	$\mathcal{O}(\lvert x\rvert^{-1})$	$\mathcal{O}(\lvert x\rvert^{-2})$	$\mathcal{O}(\lvert x\rvert^{-3})$

Weak singular functions can be integrated without regularization through a coordinate transformation and regular Gaussian quadrature. Strong singularities require a Cauchy Principal Value regularization (direct approach [GG90]) or regularization by rigid body motion technique (indirect approach [Bon95a]). For hypersingular terms, Hadamards concept of finite parts (direct technique [Gui94]) or indirect methods ([GMI90]) can be used.

regularization: For the numerical treatment of singular integrals, a regularization procedure is required. The singularities of the BIE are cancelling in total, therefore it is advisable to regroup the integral terms such that no non-integrable singularities occur. The perhaps simplest method for this is obtained by converting the free term to a rigid mode integral (regularization by rigid motion). For a more detailed study, see the review paper of Tanaka et al. [TSS94].

Appendix B

Special Distributions

Definition B.1 The Dirac-distribution
The n-dimensional Dirac-distribution $\delta(x) = \prod_{k=1}^{n} \delta(x_k)$ is defined by

$$\begin{aligned} \int_{\mathbb{R}^n} \delta(x)\,\mathrm{d}x &= 1, && x \in \mathbb{R}^n; \\ \delta(x) &= 0 \quad \text{for all} && |x| \neq 0. \end{aligned} \tag{B.1}$$

This means that $\delta(x)$ is concentrated on a single point $x = 0$ with infinite density $\delta(0) \to \infty$ but finite measure $\int \delta\,\mathrm{d}x = 1$. It is the identity object concerning convolution

$$u = u * \delta \;=\; \int_{\mathbb{R}^n} u(y)\delta(x-y)\,\mathrm{d}y, \qquad x, y \in \mathbb{R}^n, \tag{B.2}$$

and its Fourier transform is

$$\delta(x) \quad \overset{\mathcal{F}}{\leftrightarrow} \quad 1, \qquad x \in \mathbb{R}^n. \tag{B.3}$$

The Fourier transform of the translated Dirac-distribution δ is even defined for complex $y \in \mathbb{C}^n$, cf [Cha87],

$$\delta(x-y) \quad \overset{\mathcal{F}}{\leftrightarrow} \quad e^{-i<y,\hat{x}>}, \qquad x, y \in \mathbb{C}^n. \tag{B.4}$$

All differentiations can be expressed by a convolution with $\delta^\alpha = \partial^\alpha \delta$

$$u * (\partial^\alpha \delta) \;=\; \int_{\mathbb{R}^n} u(y)\partial^\alpha \delta(x-y)\,\mathrm{d}y = \partial^\alpha u(x), \qquad x, y \in \mathbb{R}^n. \tag{B.5}$$

The product with another distribution u can be simplified by the following relation[1]

$$u(x)\,\partial_x^k\delta(x-y) \;=\; \sum_{l=0}^{k}\binom{k}{k-l}(-1)^l\partial^l u(y)\,\partial_x^{k-l}\delta(x-y). \quad (B.6)$$

If a boundary $\partial\Omega$ is described by $\psi(x)=0$ with a function $\psi\in C^\infty$ then the integration of a distribution u along this boundary can be described by the scalar product

$$\langle\delta(\psi),u\rangle = \int_{\psi=0} u(x)\,\mathrm{d}x. \quad (B.7)$$

The gradient $\nabla\psi$ of the hypersurface $\psi=0$ should not vanish, i.e., $\nabla\psi\neq 0$ for $\psi=0$ and the outer unit normal of this hypersurface is

$$\nu = -\frac{\nabla\psi}{|\nabla\psi|}. \quad (B.8)$$

The chain rule is still valid, we get

$$\nabla\delta(\psi) \;=\; \delta'(\psi)\nabla\psi. \quad (B.9)$$

Distributions can be represented by series which is often useful for practical computations. For the Dirac-distribution, there are many ways to construct a sequence of functions or regular distributions $\phi_k(x), k=1,2,\ldots$ which converges to the Dirac-distribution. The ϕ_k have to satisfy the following two properties, cf. [GS64], vol.1, pp.34,

1. For any $M>0$ and for $|a|\le M$ and $|b|\le M$, the quantities

$$\left|\int_a^b \phi_k(\xi)\,\mathrm{d}\xi\right| \quad (B.10)$$

   must be bounded by a constant independent of a, b, or k (in other words, depending only on M).

2. For any fixed nonvanishing a and b, we must have

$$\lim_{k\to\infty}\int_a^b \phi_k(\xi)\,\mathrm{d}\xi = \begin{cases} 0 & \text{for } a<b<0 \text{ and } 0<a<b, \\ 1 & \text{for } a<0<b. \end{cases} \quad (B.11)$$

[1]If u is singular at the same point as δ, see the theory of non-linear distributions [Col84, Col85].

One-dimensional examples for these sequences are, cf. [Kyt96, Lig66, Boi96],

$$\phi_k = \frac{1}{\varepsilon^k}\begin{cases} e^{\varepsilon^2/(x^2-\varepsilon^2)}, & |x| < \varepsilon, \\ 0, & |x| > \varepsilon; \end{cases} \tag{B.12}$$

$$\phi_k = \frac{k}{\pi}\frac{1}{1+k^2x^2};$$

$$\phi_k = \frac{k}{\sqrt{\pi}}e^{-k^2x^2};$$

$$\phi_k = \frac{1}{k\pi}\frac{\sin^2 kx}{x^2}.$$

We need these sequences for the multiplication of distributions, e.g. for $\mathrm{H}(x)\delta(x) = \delta(x)/2$ and the evaluation of the free terms in BIEs.

Definition B.2 The Heaviside-distribution
The Heaviside-distribution $\mathrm{H}(x)$ is obtained by the integration of the Dirac-distribution

$$\mathrm{H}(x) = \int_{-\infty}^{x} \delta(y)\,\mathrm{d}y = \begin{cases} 1 & \ldots \quad x > 0 \\ 0 & \ldots \quad x < 0. \end{cases} \tag{B.13}$$

In the literature, there are several definitions for the value at $x = 0$. It is in our context, i.e. for linear distributions, determined by ([Boi96])

$$\langle \mathrm{H}(x), \delta(x) \rangle = \kappa = \frac{1}{2}, \qquad x \in \mathbb{R}^1. \tag{B.14}$$

For the multi-dimensional Heaviside-distribution, we can define the cutoff distribution for a domain $\Omega \subset \mathbb{R}^n$ by (compare Chapter 3)

$$\chi(x) := \begin{cases} 1 & \ldots \quad x \in \Omega \\ \kappa(x) & \ldots \quad x \in \partial\Omega \\ 0 & \ldots \quad x \notin \bar{\Omega} = \Omega \cup \partial\Omega. \end{cases} \tag{B.15}$$

which can be expressed by

$$\chi(x) = \mathrm{H}(\psi(x)), \tag{B.16}$$

with a function $\psi \in C^\infty(\mathbb{R}^n)$ as defined for (B.7). The integration of a distribution u over the domain Ω can be described by

$$\langle \mathrm{H}(\psi), u \rangle = \int_{\psi \geq 0} u(x)\,\mathrm{d}x. \tag{B.17}$$

The value $\kappa(x)$ on the boundary $\partial\Omega$ is uniquely defined, as shown in the following example.

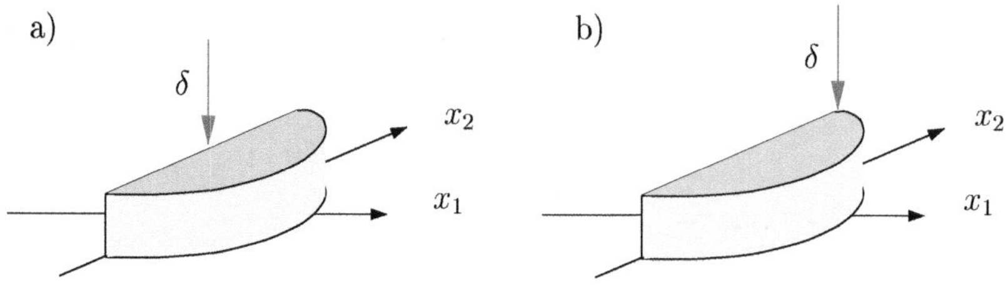

Figure B.1: *Examples for the edge and corner terms* $\kappa(x), x \in \partial\Omega$

Example B.1 The edge value κ

The value $\kappa(x)$ of the cutoff distribution χ of the semicircle

$$\chi(x) = \mathrm{H}(x_1)\mathrm{H}(1 - x_1^2 - x_2^2). \tag{B.18}$$

can be determined by exploiting the sifting property of the Dirac-distribution

$$\kappa(x) = \int_{\mathbb{R}^2} \chi(y)\delta(x - y)\,\mathrm{d}y, \qquad x \in \partial\Omega. \tag{B.19}$$

This is for the origin $x_1 = x_2 = 0$ (Fig.B.1a)

$$\begin{aligned} \kappa(0,0) &= \int_{\mathbb{R}^2} \mathrm{H}(y_1)\mathrm{H}(1 - y_1^2 - y_2^2)\delta(y_1)\delta(y_2)\,\mathrm{d}y_1\,\mathrm{d}y_2 \\ &= \int_{\mathbb{R}^2} \mathrm{H}(y_1)\delta(y_1)\delta(y_2)\,\mathrm{d}y_1\,\mathrm{d}y_2. \end{aligned}$$

The introduction of a Dirac-sequence yields

$$\kappa(0,0) = \lim_{k\to\infty} \frac{k^2}{\pi} \int_{\mathbb{R}^2} \mathrm{H}(y_1)\,\mathrm{e}^{-k^2(y_1^2+y_2^2)}\,\mathrm{d}y_1\,\mathrm{d}y_2 = \frac{1}{2}.$$

This is the standard factor $\kappa(x) = 1/2$ for smooth parts of a boundary[2]

$$\mathrm{H}(x_1)\delta(x_1) = \frac{1}{2}\delta(x_1). \tag{B.20}$$

For the corner at $x_1 = 0, x_2 = 1$ (Fig.B.1b), $\chi(y)$ is approximated locally by the tangential cutoff distribution $\mathrm{H}(y_1)\mathrm{H}(1 - y_2)$. The value $\kappa(0,1)$ is then obtained from

$$\kappa(0,1) = \lim_{k\to\infty} \frac{k^2}{\pi} \int_{\mathbb{R}^2} \mathrm{H}(y_1)\mathrm{H}(1 - y_2)\mathrm{e}^{-k^2(y_1^2+(1-y_2)^2)}\,\mathrm{d}y_1\,\mathrm{d}y_2 = \frac{1}{4},$$

[2]This value can be found in the literature of non-linear generalized functions [Boi96], p. 48, example 2.32, and [Ste95], p.47, example 2.

which is precisely the value $\kappa(x)$ known for a rectangular corner. For arbitrary angles θ (Fig. B.2), we get in $\mathbb{R}^2$

$$\kappa(x) = \lim_{k\to\infty} \frac{k^2}{\pi} \int_{\mathbb{R}^2} \mathrm{H}(y_2)\mathrm{H}(y_1 \tan\theta - y_2)\mathrm{e}^{-k^2(y_1^2+y_2^2)}\,\mathrm{d}y = \frac{\theta(x)}{2\pi}. \tag{B.21}$$

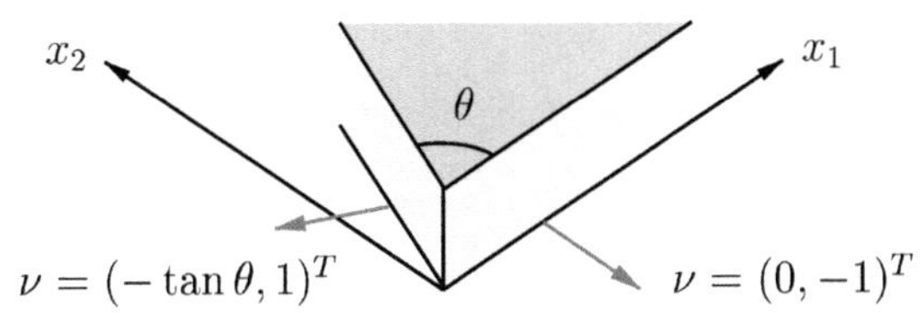

Figure B.2: *Corner part of the cutoff distribution for arbitrary angles θ*

The gradient of the cutoff distribution leads to a definition of the normal vector of the boundary even for non-smooth $\partial\Omega$.

Example B.2 The gradient of the cutoff distribution

The normal vector $\nu = -\nabla\psi = 2(x_1, x_2)^T$ of a unit circle is obtained from

$$\begin{aligned} \nabla\chi &= \nabla\mathrm{H}(1 - x_1^2 - x_2^2) = \delta(1 - x_1^2 - x_2^2)\begin{pmatrix} -2x_1 \\ -2x_2 \end{pmatrix} \\ &= \delta\left(\frac{1 - x_1^2 - x_2^2}{2|x|}\right)\frac{1}{2|x|}\begin{pmatrix} -2x_1 \\ -2x_2 \end{pmatrix}. \end{aligned}$$

For non-smooth boundaries, the cutoff distribution χ is defined by a product (or a sum of products) of Heaviside-distributions $\prod_k \mathrm{H}(\psi_k)$. Its gradient $\nabla\chi$ leads to a normal vector ν defined as for smooth boundaries.

Example B.3 The normal vector for a rectangular corner

The normal vector ν for a rectangular corner in $\mathbb{R}^2$ (Fig. B.3) is defined by the gradient of the cutoff distribution $\chi = \mathrm{H}(x_1)\mathrm{H}(x_2)$:

$$\nabla\chi = \begin{pmatrix} \delta(x_1)\mathrm{H}(x_2) \\ \mathrm{H}(x_1)\delta(x_1) \end{pmatrix} = \delta(x_1)\mathrm{H}(x_2)\begin{pmatrix} 1 \\ 0 \end{pmatrix} + \delta(x_2)\mathrm{H}(x_1)\begin{pmatrix} 0 \\ 1 \end{pmatrix}.$$

Thus, the normal vectors on the edges are $\nu = (-1, 0)^T$ for $x_1 = 0$ and $\nu = (0, -1)^T$ for $x_2 = 0$. Note that for the first differentiation, there is no special term at the corner $x_1 = x_2 = 0$.

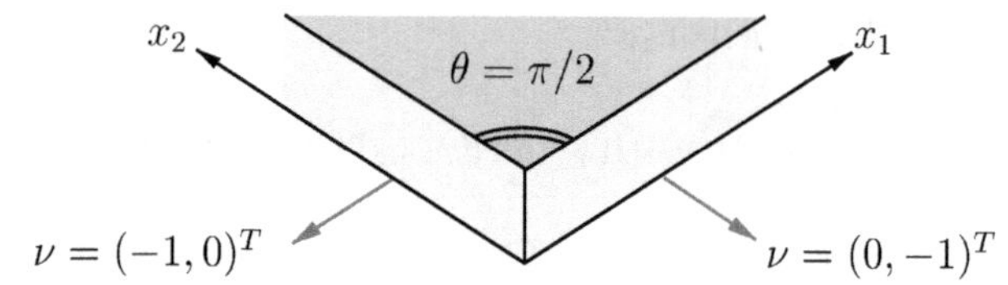

Figure B.3: *Cutoff distribution of rectangular corner and corresponding normal vectors ν defined by $\nabla\chi$*

Example B.4 The normal vector for corners with arbitrary angles
An arbitrary angle θ in $\mathbb{R}^2$ is described by the cutoff distribution $\chi = \mathrm{H}(x_2)\mathrm{H}(x_1 \tan\theta - x_2)$. The gradient leads to

$$\nabla\chi = \begin{pmatrix} \mathrm{H}(x_2)\delta(x_1 \tan\theta - x_2)\tan\theta \\ \delta(x_2)\mathrm{H}(x_1 \tan\theta - x_2) - \mathrm{H}(x_2)\delta(x_1 \tan\theta - x_2) \end{pmatrix}.$$

Thus, the normal vector is $\nu = (0, -1)^T$ for the edge $x_2 = 0$ and $\nu = (-\tan\theta, 1)$ for $x_2 = x_1 \tan\theta$.

Example B.5 The gradient of χ for the semicircle
The gradient of the semicircle (cf. Example B.1) is

$$\begin{aligned} \nabla\chi &= \nabla\{\mathrm{H}(x_1)\mathrm{H}(1 - x_1^2 - x_2^2) \\ &= \begin{pmatrix} \delta(x_1)\mathrm{H}(1 - x_1^2 - x_2^2) + \mathrm{H}(x_1)\delta(1 - x_1^2 - x_2^2)(-2x_1) \\ \mathrm{H}(x_1)\delta(1 - x_1^2 - x_2^2)(-2x_2) \end{pmatrix}. \end{aligned}$$

The normal vector is hence $\nu = (-1, 0)^T$ for the straight part of the boundary $x_1 = 0$ and $\nu = (-2x_1, -2x_2)^T$ for the curved part.

For differential operators of higher degree (higher than two), additional derivatives of the cutoff distribution χ are required. They lead to additional corner terms which is shown in the next example.

Example B.6 Higher derivatives of the cutoff distribution
The second derivatives of the rectangular corner of Example B.3 with $\chi = \mathrm{H}(x_1)\mathrm{H}(x_2)$ are

$$\begin{bmatrix} \partial_{11} & \partial_{12} \\ \partial_{21} & \partial_{22} \end{bmatrix} \chi = \begin{bmatrix} \delta'(x_1)\mathrm{H}(x_2) & \delta(x_1)\delta(x_2) \\ \delta(x_1)\delta(x_2) & \delta'(x_2)\mathrm{H}(x_1) \end{bmatrix}.$$

Thus, we get for the mixed derivative $\partial_{12}\chi$ particular corner terms $\delta(x_1)\delta(x_2)$. For an arbitrary angle with $\chi = \mathrm{H}(x_2)\mathrm{H}(x_1 \tan\theta - x_2)$, we get as second differentiations

$$\begin{aligned}
\partial_{11}\chi &= \mathrm{H}(x_2)\delta'(x_1 \tan\theta - x_2)\tan^2\theta, \\
\partial_{12}\chi &= \delta(x_2)\delta(x_1 \tan\theta - x_2)\tan\theta - \mathrm{H}(x_2)\delta'(x_1 \tan\theta - x_2)\tan\theta, \\
\partial_{21}\chi &= \partial_{12}\chi, \\
\partial_{22}\chi &= \delta'(x_2)\mathrm{H}(x_1 \tan\theta - x_2) - 2\delta(x_2)\delta(x_1 \tan\theta - x_2) + \\
&+ \mathrm{H}(x_2)\delta'(x_1 \tan\theta - x_2).
\end{aligned}$$

Once more, we obtained additional corner terms.

Appendix C

Integration of BEM matrices

C.1 Analytical integrations

C.1.1 Singular integrals by Fourier transform

A large number of the one- and two-dimensional integrations can be computed analytically by means of the symbolic computation program MAPLE [Wat97]. For this, some additional tools were developed based on the routine for the analytical Fourier transformation. The main problem with this was the treatment of the singularities. The integration tools implemented in MAPLE are in general not able to handle strong and hypersingular integrals. The Fourier transform is based on distribution theory and therefore by far better suited for singular integrals.

The n-dimensional integrations in Fourier space of the kernels $\hat{f}$ are computed by successively applying the following formula for $\hat{x}_k, k = 1 \ldots n$

$$\int_{\mathbb{R}^1} \hat{f}(\hat{x}_k)\,\mathrm{d}\hat{x}_k \;=\; \lim_{x_k \to 0} \mathcal{F}\{\hat{f}(\hat{x}_k)\} = \lim_{x \to 0} \int_{\mathbb{R}^1} \hat{f}(\hat{x}_k)\,\mathrm{e}^{i<\hat{x}_k, x_k>}\,\mathrm{d}\hat{x}_k. \quad \text{(C.1)}$$

C.1.2 Additional regular Fourier pairs

In the MAPLE version used for this thesis, the analytical Fourier transform is based on a table of predefined pairs of Fourier transform. Because they are in general not sufficient for the integrations required here, additional pairs have to be defined which can be easily done by the MAPLE routine *addtable*.

For example, the Fourier transform of the exponential function is added to the Fourier table with

$$e^{c_1 x_k} \quad \overset{\mathcal{F}}{\leftrightarrow} \quad 2\pi\delta(\hat{x}_k + ic_1), \qquad c_1 \in \mathbb{C}, i = \sqrt{-1}.$$

The complex Dirac-distribution is a ultradistribution, which is not implemented originally. Similarly, the following pairs are added (see Lighthill [Lig66] for the justification) :

$$\frac{e^{c_1 x_k}}{x_k^m} \quad \overset{\mathcal{F}}{\leftrightarrow} \quad -i\pi \frac{(-i\hat{x}_k + c_1)^{m-1}}{(m-1)!} \operatorname{csgn}(\hat{x}_k + ic_1), \qquad m \in \mathbb{N}.$$

csgn is the complex signum function.

C.1.3 Additional singular Fourier pairs

Lighthill gives as transform of $\operatorname{sgn} x_k / x_k^m$, with an integer $m > 0$ the following result :

$$\frac{\operatorname{sgn} x_k}{x_k^m} \quad \overset{\mathcal{F}}{\leftrightarrow} \quad -2\frac{(-i\hat{x}_k)^{m-1}}{(m-1)!}(\ln|x_k| + C). \tag{C.2}$$

The constant C occurs because the functions $\operatorname{sgn} x_k / x_k^m$ are determined only up to an arbitrary multiple of the Dirac-distribution $\delta^{(m-1)}$ at the singular point $x_k = 0$, see Lighthill [Lig66], p.57. This constant C has to be defined for each fundamental solution which is shown for the example of the Poisson equation in the following.

Example C.1.1 Poisson equation
The integration of the Fourier fundamental solution $\hat{U}$ should yield the value of the fundamental solution $U(0)$ at $x = 0$ in the original space. We have for the Laplacian in $\mathbb{R}^2$:

$$\hat{U} = -\frac{1}{\hat{x}_1^2 + \hat{x}_2^2}.$$

The first integration over $\hat{x}_2 = -\infty \ldots \infty$ yields

$$-\frac{1}{(2\pi)^2}\int_{\mathbb{R}^1} \frac{1}{\hat{x}_1^2 + \hat{x}_2^2}\,\mathrm{d}\hat{x}_2 = -\frac{1}{2\pi}\frac{\operatorname{sgn}\hat{x}_1}{2\hat{x}_1}.$$

Based on (C.2) with $m = 1$, we get for the integration with respect to $\hat{x}_1$ the following divergent expression

$$-\frac{1}{2\pi}\int_{\mathbb{R}^1} \frac{\operatorname{sgn}\hat{x}_1}{2\hat{x}_1}\,\mathrm{d}\hat{x}_1 = -\frac{1}{2\pi}\lim_{x_1\to 0}(\ln|x_1| + C).$$

The known fundamental solution in the original space is already given in equation (2.5)

$$U(x) = -\frac{1}{2\pi}(\ln|\sqrt{x_1^2 + x_2^2}| + C_0).$$

The limit process $x_k \to 0, k = 1, 2$ *yields*

$$-\frac{1}{2\pi}\lim_{x_1 \to 0}(\ln|x_1| + C_0).$$

Thus it can be seen that the two constants C *and* C_0 *are the same. It is well known that the constant* C_0 *in the original space can be chosen arbitrarily. It expresses additional homogeneous solutions which can be added to the fundamental solution without changing the solution. In general it is chosen to be zero* $C = C_0 = 0$*. Hence we have established the following Fourier pair for the MAPLE table*

$$\frac{\operatorname{sgn} x_k}{x_k} \overset{\mathcal{F}}{\leftrightarrow} -2(\ln|x_k|).$$

The table entries for $m > 1$ *are obtained in a similar manner (Tab.C.1.1). For the Fourier pendant to* $\mathrm{e}^{c_1 \hat{x}_k} \operatorname{sgn} x_k / x_k, c_1 \in \mathbb{C}$ *the table Tab.C.1.1 can be*

original space	$\overset{\mathcal{F}}{\leftrightarrow}$	Fourier space
$\operatorname{sgn} x_k / x_k$	$\overset{\mathcal{F}}{\leftrightarrow}$	$-2(\ln|x_k|)$
$\operatorname{sgn} x_k^2 / x_k^2$	$\overset{\mathcal{F}}{\leftrightarrow}$	$2i\hat{x}_k(\ln|\hat{x}_k| - 1)$
$\operatorname{sgn} x_k^3 / x_k^3$	$\overset{\mathcal{F}}{\leftrightarrow}$	$2\hat{x}_k^2/2!(\ln|\hat{x}_k| - 3/2)$
$\operatorname{sgn} x_k^4 / x_k^4$	$\overset{\mathcal{F}}{\leftrightarrow}$	$-2i\hat{x}_k^3/3!(\ln|\hat{x}_k| - 11/6)$
$\operatorname{sgn} x_k^5 / x_k^5$	$\overset{\mathcal{F}}{\leftrightarrow}$	$-2\hat{x}_k^4/4!(\ln|\hat{x}_k| - 25/12)$

Table C.1: Fourier pairs for $\operatorname{sgn} x_k / x_k^m$

used by substituting $\hat{x}_k$ *by* $\hat{x}_k + ic_1$*. Note that then* $\ln|\hat{x}_k|$ *should be replaced by the complex logarithmic function*

$$\ln|\hat{x}_k| \;\to\; \ln(\hat{x}_k \operatorname{csgn} \hat{x}_k).$$

C.2 Numerical integrations

As an example for the numerical integrations, the computation of the following matrix entries for the stationary heat conduction equation are investigated (see (5.3)

$$H^{ji} = \frac{1}{(2\pi)^2} \left\langle \hat{\phi}_{\rm t}^j(-\hat{x}), \hat{\phi}_{\rm t}^i \hat{U} \right\rangle .$$

All integrals have inherited their singularity from the fundamental solution $\hat{U}$. We compute first the off-diagonal entries by subtracting from them the diagonal term $\tilde{H}^{ji} = H^{ji} - \lambda^{ji} H^{jj}$ for all $j \neq i$

$$\tilde{H}^{ji} = \frac{1}{(2\pi)^2} \left\langle \hat{\phi}_{\rm t}^j(-\hat{x}), (\hat{\phi}_{\rm t}^i - \lambda^{ji} \hat{\phi}_{\rm t}^j) \hat{U} \right\rangle .$$

$\lambda^{ji} = L_{\rm e}^i / L_{\rm e}^j$ is the ratio between the length of the i-th element and the length of element j. These integral are regular and can be computed numerically.

The diagonal kernels are simpler than the off-diagonal kernels because they are not composed by a trial and test function which have different orientation. They are computed analytically as shown in the previous section. The final matrices are obtained by

$$H^{ji} = \tilde{H}^{ji} + \lambda^{ji} H^{jj} .$$

Bibliography

[AAH89] A. Abdel-Akher and G.A. Hartley. Evaluation of boundary integrals for plate bending. *Int. J. Numer. Meth. Engrg.*, 28:75–93, 1989.

[AAH90] A. Abdel-Akher and G.A. Hartley. An integration procedure for curved boundary elements. *Comm. Appl. Numer. Meth.*, 6:49–56, 1990.

[AB90] S. Ahmad and P.K. Banerjee. Inelastic transient dynamic analysis of three-dimensional problems by BEM. *Int. J. Numer. Meth. Engrg.*, 29:371–390, 1990.

[ABC+00] A. Averbuch, E. Braverman, R. Coifman, M. Israeli, and A. Sidi. Efficient computation of oscillatory integrals via adaptive multiscale local Fourier bases. *Appl. and Comput. Harmonic Anal.*, 9:19–53, 2000.

[Ada76] R.A. Adams. *Sobolev Spaces.* Academic Press, New York, 1976.

[AHFEZ99] K. Al-Hosani, S. Fadhil, and A. El-Zafrany. Fundamental solution and boundary element analysis of thick plates in Winkler foundation. *Comput. Struct.*, 70:325–336, 1999.

[Ali97] M.H. Aliabadi. Boundary element formulations in fracture mechanics. *ASME - Appl. Mech. Review*, 50(2):83–96, 1997.

[Ant88] H. Antes. *Anwendungen der Methode der Randelemente in der Elastodynamik und der Fluiddynamik.* Teubner, Stuttgart, 1988.

[AP92] H. Antes and P.D. Panagiotopoulos. *The Boundary Integral Approach to Static and Dynamic Contact Problems.* Birkhäuser, Basel, 1992.

[Ara94] F.C. Araújo. *Zeitbereichslösung linearer dreidimensionaler Probleme der Elastodynamik mit einer gekoppelten BE/FE-Methode.* PhD thesis, Technische Universität Braunschweig, 1994.

[AS97] H. Andrä and E. Schnack. Integration of singular Galerkin-type boundary element integrals for 3D elasticity problems. *Numer. Math.*, 76(2):143–165, 1997.

[AW83] D.N. Arnold and W.L. Wendland. On the asymptotic convergence of collocation methods. *Math. Comput.*, 41:349–381, 1983.

[BA93] C.A. Brebbia and M.H. Aliabadi, editors. *Adaptive Finite and Boundary Element Methods.* Computational Mechanics Publications, Southhampton, 1993.

[Baa99] J. Baaran. *Schallfeldanalyse bei sich bewegenden schallerzeugenden Körpern.* PhD thesis, Technische Universität Braunschweig, 1999.

[Ban94] P.K. Banerjee. *The boundary element methods in engineering.* McGraw-Hill, London, 2nd edition, 1994.

[Bar72] D.M. Barnett. A precise evaluation of derivatives of the anisotropic elastic Green's functions. *Phys. Stat. Sol. (b)*, 49:741–748, 1972.

[Bar99] D. Barbier. *Méthode des potentiels retardés pour la simulation de la diffraction d'onde élastodynamique par une fissure tridimensionnelle.* PhD thesis, Ecole Polytechnique, Palaiseau, 1999.

[BB81] P.K. Banerjee and R. Butterfield. *Boundary Element Methods in Engineering Science.* McGraw-Hill, London, 1981.

[BCR91] G. Beylkin, R. Coifman, and V. Rokhlin. Fast wavelet transforms and numerical algorithms i. *Comm. Pure and Appl. Math.*, 44:141–183, 1991.

[Bec92] A.A. Becker. *The Boundary Element Method in Engineering: a complete course.* McGraw-Hill, New York, 1992.

[Bes87] D.E. Beskos. Boundary element methods in dynamic analysis. *ASME - Appl. Mech. Review*, 40:1–23, 1987.

[Bes91] D.E. Beskos, editor. *Boundary Element Analysis of Plates and Shells.* Springer, Berlin, 1991.

[Bes97] D.E. Beskos. Boundary element methods in dynamic analysis: Part II (1986–1996). *ASME - Appl. Mech. Review*, 50(3):149–197, 1997.

[Bia90] H.A. Biagioni. *A Nonlinear Theory of Generalized Functions.*, volume 1421 of *Lecture Notes in Mathematics.* Springer, Berlin, 1990.

[BK83] R. Bausinger and G. Kuhn. *Die Boundary-Element-Methode.* Expert-Verlag, Ehingen, 1983.

[BK92] P.K. Banerjee and S. Kobayashi, editors. *Advanced Dynamic Analysis by Boundary Element Methods.* Number 7 in Developments in Boundary Element Methods. Elsevier, London, 1992.

[Bla98] C. Blatter. *Wavelets - Eine Einführung.* Vieweg, Braunschweig, 1998.

[BM96] M. Bonnet and S. Mukherjee. Implicit BEM formulations for usual and sensitivity problems in elasto-plasticity using the consistent tangent operator concept. *Int. J. Solids Struct.*, 33(30):4461–4480, 1996.

[BMP98] M. Bonnet, G. Maier, and C. Polizzotto. Symmetric Galerkin boundary element methods. *ASME - Appl. Mech. Review*, 51(11):669–704, 1998.

[BNN89] E. Becache, J.-C. Nedelec, and N. Nishimura. Regularization in 3D for anisotropic elastodynamic crack and obstacle problems. Technical Report 205, Ecole Polytechnique, Centre de Mathématiques Appliquées, Palaiseau, 1989.

[Boi96] V. Boie. *Partielle Multiplikationen in Distributionenräumen.* PhD thesis, Universität Hannover, Institut für Angewandte Mathematik, 1996.

[Bon86] M. Bonnet. *Méthode des équations intégrales régularisées en élastodynamique tridimensionnelle.* PhD thesis, Ecole Nationale des Ponts et Chaussées, Paris, 1986.

[Bon95a] M. Bonnet. *Équations intégrales et éléments de frontière.* CNRS Éditions / Eyrolles, Paris, 1995.

[Bon95b] M. Bonnet. Regularized direct and indirect symmetric variational BIE formulations for three-dimensional elasticity. *Engrg. Anal. Bound. Elem.*, 15:93–102, 1995.

[Bon97] M. Bonnet. Differentiability of strongly singular and hypersingular boundary integral formulations with respect to boundary perturbations. *Comput. Mech.*, 19:240–246, 1997.

[Bon99] M. Bonnet. *Boundary Integral Equation Methods for Solids and Fluids.* Wiley, New York, 1999.

[BP92] C.A. Brebbia and P.W. Partridge. *Boundary Elements in Fluid Dynamics.* Kluwer, Dordrecht, 1992.

[BR86] P.K. Banerjee and S.T. Raveendra. Advanced boundary element analysis of two- and three-dimensional problems of elasto-plasticity. *Int. J. Numer. Meth. Engrg.*, 23:985–1002, 1986.

[Bra86] R.N. Bracewell. *The Fourier Transform and its Applications.* McGraw-Hill, New York, 1986.

[Bre78] C.A. Brebbia. *The Boundary Element Method for Engineers.* Pentech Press, London, 1978.

[Bre84] C.A. Brebbia. *Boundary Element Techniques in Computer-Aided Engineering.* Kluwer, Dordrecht, 1984.

[BSS89] J. Balaš, J. Sládek, and V. Sládek. *Stress Analysis by Boundary Element Methods.* Elsevier, Amsterdam, 1989.

[BSW99] M. Bonnet, A.M. Sändig, and W.L. Wendland, editors. *Mathematical Aspects of Boundary Element Methods.* Research Notes in Mathematics. Chapman and Hall, Dordrecht, 1999.

[BTW84] C.A. Brebbia, J.C.F. Telles, and L.C. Wrobel. *Boundary Element Techniques.* Springer, Berlin, 1984.

[Bud93] G. Budzick. *Integralsätze für Schwartzsche Distributionen.* PhD thesis, Technische Hochschule Darmstadt, 1993.

[Bui78] H.D. Bui. Some remarks about the formulation of three-dimensional thermoelasto-plastic problems by integral equations. *Int. J. Solids Struct.*, 14:935–939, 1978.

[BV94] D.M. Bond and S.A. Vavasis. Fast wavelet transforms for matrices arising from boundary element methods. Technical report, Cornell University, Ithaca, 1994.

[BW92] G. Beer and J.O. Watson. *Introduction to Finite and Boundary Element Methods for Engineers.* Wiley, Chichester, 1992.

[CB91] R.D. Ciskowski and C.A. Brebbia. *Boundary Element Methods in Acoustics.* Kluwer, Dordrecht, 1991.

[CDMZ99] A. Carini, M. Diligenti, P. Maranesi, and M. Zanella. Analytical integration for two-dimensional elastic analysis by the symmetric Galerkin boundary element method. *Comput. Mech.*, 23:308–323, 1999.

[CG97] C. Carstensen and J. Gwinner. FEM and BEM coupling for a nonlinear transmission problem with Signorini contact. *SIAM - J. Numer. Anal.*, 34(5):1845–1864, 1997.

[Cha87] D.C. Champeney. *Handbook of Fourier Theorems.* Cambridge University Press, 1987.

[Chu92] C.K. Chui. *An Introduction to Wavelets.* Academic Press, Boston, 1992.

[CL95] T. Chen and F.Z. Lin. Boundary integral formulations for three-dimensional anisotropic piezoelectric solids. *Comput. Mech.*, 15:485–496, 1995.

[Cle98] D.L. Clements. Fundamental solutions for second order linear elliptic partial differential equations. *Comput. Mech.*, 22:26–31, 1998.

[Clo90] D. Clouteau. *Propagation d'ondes dans les milieux hétérogènes. Application à la tenue des ouvrages sous séismes.* PhD thesis, Ecole Centrale de Paris, 1990.

[Col84] J.F. Colombeau. *New Generalized Functions and Multiplication of Distributions*, volume 84. North Holland Mathematics Studies, Amsterdam, 1984.

[Col85] J.F. Colombeau. *Elementary Introduction to New Generalized Functions*, volume 113. North Holland Mathematics Studies, Amsterdam, 1985.

[Col92] J.F. Colombeau. *Multiplication of Distributions*, volume 1532. Springer Lecture Notes in Mathematics, Heidelberg, 1992.

[Con74] F. Constantinescu. *Distributionen und ihre Anwendung in der Physik.* Teubner, Stuttgart, 1974.

[Cos84] M. Costabel. *Starke Elliptizität von Randintegraloperatoren erster Art.* Habiliation, Technische Hochschule Darmstadt, 1984.

[Cos88] M. Costabel. Boundary integral operators on Lipschitz domains: elementary results. *SIAM - J. Math. Anal.*, 19:613–626, 1988.

[CR83] D.L. Clements and C. Rogers. A boundary integral equation for the solution of a class of problems in anisotropic inhomogeneous thermostatics and elastostatics. *Quart. Appl. Math.*, 41:90–105, 1983.

[CR96] T.A. Cruse and J.D. Richardson. Non-singular Somigliana stress identities in elasticity. *Int. J. Numer. Meth. Engrg.*, 39:3273–3304, 1996.

[Cru88] T.A. Cruse. *Boundary Element Analysis in Computational Fracture Mechanics.* Kluwer, Dordrecht, 1988.

[CS83] S.L. Crouch and A.M. Starfield. *Boundary Element Methods in Solid Mechanics.* George Allen & Unwin Ltd., London, 1983.

[CS99] W. Chen and T. Shioya. Green's function of an external circular crack in a transversely isotropic piezoelectric medium. *JSME Int. J.*, 42(1):73–79, 1999.

[CX98] Z. Chen and Y. Xu. The Petrov-Galerkin and iterated Petrov-Galerkin methods for second-kind integral equations. *SIAM - J. Numer. Anal.*, 35(1):406–434, 1998.

[CZ92] G. Chen and J. Zhou. *Boundary Element Methods.* Academic Press, London, 1992.

[DA00] C.H. Daros and H. Antes. Dynamic fundamental solutions for transversely isotropic piezoelectric materials of crystal class 6 mm. *Int. J. Solids Struct.*, 37(11):1639–1658, 2000.

[Dal88] R. Dallmann. *Kopplung von Rand- und Finite-Element-Methode beim Tunnelvortrieb im Voll- und Halbraum.* PhD thesis, Technische Universität Braunschweig, 1988.

[Dal93] R. Dallner. *Behandlung inelastischer 3D-Probleme mittels der Randelementmethode.* Fortschritt-Berichte 18/126, Universität Erlangen-Nürnberg, VDI-Verlag, Düsseldorf, 1993.

[Dan89] P. Dangla. *Couplage éléments finis - équations intégrales en élastodynamique et interaction sol-structure.* PhD thesis, Ecole Nationale des Ponts et Chaussées, Paris, 1989.

[Dar99] C.H. Daros. *Wave propagation in unbounded piezoelectric media of transversely isotropic symmetry.* PhD thesis, Technische Universität Braunschweig, 1999.

[Dau92] I. Daubechies. *Ten Lectures on Wavelets.* SIAM Publications, Philadelphia, 1992.

[Dav91] G. David. *Wavelets and Singular Integrals on Curves and Surfaces.* Number 1465 in Lecture Notes in Mathematics. Springer, Berlin, 1991.

[Deb96] A. Deb. Boundary element analysis of anisotropic bodies under thermomechanical body force loadings. *Comput. Struct.*, 58(4):715–726, 1996.

[DG00] F.M.E. Duddeck and M. Geisenhofer. A general boundary element method for anisotropic plates. In *Symp. of the Int. Assoc. for Boundary Element Methods (IABEM)*, pages 81–83, Brescia, 2000.

[DG01] F.M.E. Duddeck and M. Geisenhofer. A general boundary element method for anisotropic plates. *Comput. Mech.*, 2001. to be published.

[DHW91] A. Deb, D.P. Henry, and R.B. Wilson. Alternate BEM formulations for 2- and 3-D anisotropic thermoelasticity. *Int. J. Solids Struct.*, 27(13):1721–1738, 1991.

[Die78] J. Dieudonné. *Treatise on Analysis.* Academic Press, London, 1978.

[DL69] P.H. Dederichs and G. Leibfried. Elastic Green's function for anisotropic cubic crystals. *Phys. Rev.*, 188(3):1175–1183, 1969.

[DL99] H. Ding and J. Liang. The fundamental solutions for transversely isotropic piezoelectricity and boundary element method. *Comput. Struct.*, 71:447–455, 1999.

[DL00] R. Dautray and J.-L. Lions. *Mathematical Analysis and Numerical Methods for Science and Technology*, volume I-VI. Springer, Berlin, 2000.

[Dom93] J. Dominguez. *Boundary Elements in Dynamics.* Computational Mechanics Publications, Southhampton, 1993.

[DP01] F.M.E. Duddeck and A. Pomp. Calculation of BEM matrices by Fourier transform. *AMS - Math. of Computation*, 2001. to be published.

[DPCK98] A. Druzhinin, H. Pedersen, M. Campillo, and W. Kim. Elastic Kirchhoff-Helmholtz synthetic seismograms. *Pure Appl. Geophys.*, 151:17–45, 1998.

[DPS93] W. Dahmen, S. Prößdorf, and R. Schneider. Wavelet approximation methods for pseudodifferential equations. II: Matrix compression and fast solution. *Advances Comput. Math.*, 1:259–335, 1993.

[Dre82] W. Drexler. *Ein Beitrag zur Lösung rotationssymmetrischer, thermoelastischer Kerbprobleme mittels der Randintegralgleichungsmethode.* PhD thesis, Technische Universität München, Munich, 1982.

[dRS94] J.J. do Rêgo Silva. *Acoustic and Elastic Wave Scattering using Boundary Elements.* Computational Mechanics Publications, Southampton, 1994.

[Dud97a] F.M.E. Duddeck. Fourier transform in bounded domains. *Meccanica*, 32(3):197–204, 1997.

[Dud97b] F.M.E. Duddeck. *Functional Analysis in Solid Mechanics - Spatial and Temporal Fourier Transform of Energy Methods (in German).* PhD thesis, Technische Universität München, Munich, 1997.

[Dud99a] F.M.E. Duddeck. Alternative forms of boundary integral equations for viscous media. In Fryba & Náprstek, editor, *Structural Dynamics*, volume 1, pages 97–101, Rotterdam, 1999. EURODYN '99, Balkema.

[Dud99b] F.M.E. Duddeck. A boundary element method for anisotropic media. In W. Wunderlich, editor, *Europ. Conf. Comput. Mech. (ECCM)*, pages 1–10, Munich, 1999. (CD–ROM).

[Dud00a] F.M.E. Duddeck. Analytic integration of bi- and tri-dimensional straight boundary elements. In *Symp. of the Int. Assoc. for Boundary Element Methods (IABEM)*, pages 77–79, Brescia, 2000.

[Dud00b] F.M.E. Duddeck. A general boundary element method for homogeneous differential operators - linear or non-linear. In *Europ. Congr. Comput. Meth. Appl. Sci. Engrg. (ECCOMAS)*, Barcelona, 2000. CD–ROM.

[Dud01a] F.M.E. Duddeck. Fourier-BEM - or what to do if no fundamental solution is available ? *Meccanica*, 2001. to be published.

[Dud01b] F.M.E. Duddeck. Fourier transformed boundary integral equations for transient problems of elasticity and thermo-elasticity. In *First M.I.T. Conf. on Comput. Fluid and Solid Mech.*, Cambridge, MA, 2001. to be published.

[Dud01c] F.M.E. Duddeck. A Galerkin time discretization of Fourier transformed boundary integral equations. In *Europ. Conf. Comput. Mech. (ECCM)*, Krakau, 2001. to be published.

[DW96] M.L. Dunn and H.A. Wienecke. Green's functions for transversely isotropic piezoelectric solids. *Int. J. Solids Struct.*, 33(30):4571–4581, 1996.

[Elz91] A. Elzein. *Plate Stability by Boundary Element Method.* Springer, Berlin, 1991.

[ES75] A.C. Eringen and E.S. Suhubi. *Elastodynamics*, volume 2. Academic Press, New York, 1975.

[Est86] O.v. Estorff. *Zur Berechnung der dynamischen Wechselwirkung zwischen Bauwerken und ihrer Umgebung mittels zeitabhängiger Randintegralgleichungen.* PhD thesis, Ruhr-Universität Bochum, 1986.

[Fae93] B. Faermann. *Lokale a-posteriori-Fehlerschätzer bei der Diskretisierung von Randintegralgleichungen.* PhD thesis, Universität Kiel, 1993.

[FB98] A. Frangi and M. Bonnet. A Galerkin symmetric and direct BIE method for Kirchhoff elastic plates: Formulation and implementation. *Int. J. Numer. Meth. Engrg.*, 41:337–369, 1998.

[FG01] A. Frangi and M. Guiggiani. Free terms and compatibility conditions for 3D hypersingular boundary integral equations. *Z. Angewandte Math. Mech.*, 2001. to be published.

[Fie93] C. Fiedler. *Über die Berechnung von Feldgrößen der Elastostatik und Elastodynamik mit der Randelementmethode.* PhD thesis, Universität der Bundeswehr Hamburg, 1993.

[FN98] A. Frangi and G. Novati. Regularized symmetric Galerkin BIE formulations in the Laplace transform domain for 2D problems. *Comput. Mech.*, 22:50–60, 1998.

[Fra99] A. Frangi. Elastodynamics by BEM: A new formulation. *Int. J. Numer. Meth. Engrg.*, 45:721–740, 1999.

[GD96] R. Gallego and J. Domínguez. Hypersingular BEM for transient elastodynamics. *Int. J. Numer. Meth. Engrg.*, 39:1681–1705, 1996.

[GF97] L. Gaul and C. Fiedler. *Methode der Randelemente in Statik und Dynamik.* Vieweg, Braunschweig, 1997.

[GG90] M. Guiggiani and A. Gigante. A general algorithm for multi-dimensional Cauchy principal value integrals in the boundary element method. *ASME - J. Appl. Mech.*, 57:906–915, 1990.

[GGK96] L.J. Gray, D. Ghosh, and T. Kaplan. Evaluation of the anisotropic Green's function in three dimensional elasticity. *Comput. Mech.*, 17:255–261, 1996.

[GH95] G.N. Gatica and G.C. Hsiao. *Boundary-field equation methods for a class of nonlinear problems.* Pitman research notes in mathematics. Longman, Harlow, UK, 1995.

[GKRR92] M. Guiggiani, G. Krishnasamy, T.J. Rudolphi, and F.J. Rizzo. A general algorithm for the numerical solution of hypersingular boundary integral equations. *ASME - J. Appl. Mech.*, 59:604–614, 1992.

[GL65] W. Gröbner and P. Lesky. *Mathematische Methoden der Physik II.* Bibliographisches Institut, Mannheim, 1965.

[GLB00] S. Ganguly, J.B. Layton, and C. Balakrishna. Symmetric coupling of multi-zone curved Galerkin boundary elements with finite elements in elasticity. *Int. J. Numer. Meth. Engrg.*, 48:633–654, 2000.

[GM86] S. Guoshu and S. Mukherjee. Boundary element method analysis of bending of elastic plates of arbitrary shape with general boundary conditions. *Engrg. Anal. Bound. Elem.*, 3:2049–2069, 1986.

[GM92] M.G. Garroni and J.L. Menaldi. *Green's functions for second order parabolic integro-differential problems.* Pitman research notes in mathematics. Longman, Harlow, UK, 1992.

[GMI90] L.J. Gray, L.F. Martha, and A.R. Ingraffea. Hypersingular integrals in boundary element fracture analysis. *Int. J. Numer. Meth. Engrg.*, 29:1135–1158, 1990.

[GP97] L.J. Gray and G.H. Paulino. Symmetric Galerkin boundary integral fracture analysis for plane orthotropic elasticity. *Comput. Mech.*, 20:26–33, 1997.

[Gra54] D. Graffi. Über den Reziprozitätssatz in der Dynamik elastischer Körper. *Ingenieurarchiv*, 22:45–46, 1954.

[GS64] I.M. Gel'fand and G.E. Shilov. *Generalized Functions*, volume I-V. Academic Press, New York, 1964.

[GS93] J. Gwinner and E.P. Stephan. A boundary element procedure for contact problems in plane linear elastostatics. *RAIRO Math. Modell. Numer. Anal.*, 27:457–480, 1993.

[Gui94] M. Guiggiani. Hypersingular formulation for boundary stress evaluation. *Engrg. Anal. Bound. Elem.*, 13:169–179, 1994.

[GW97] G.N. Gatica and W.L. Wendland. Coupling of mixed finite elements and boundary elements for a hyperelastic interface problem. *SIAM - J. Numer. Anal.*, 34(6):2335–2356, 1997.

[GWW97] L. Gaul, M. Wagner, and W. Wenzel. Efficient field point evaluation by combined direct and hybrid boundary element methods. Bericht 97/49, Universität Stuttgart, 1997. Sonderforschungsbereich 404.

[GZ60] A.E. Green and W. Zerna. *Theoretical Elasticity.* Oxford University Press, 1960.

[Hac97] W. Hackbusch. *Integralgleichungen - Theorie und Numerik.* Teubner, Stuttgart, 1997.

[Hal94] W.S. Hall. *The Boundary Element Method.* Kluwer, Dordrecht, 1994.

[Har87] F. Hartmann. *Methode der Randelemente. Boundary Elements in der Mechanik auf dem PC.* Springer, Berlin, 1987.

[Har89] F. Hartmann. *Introduction to Boundary Elements.* Springer, Berlin, 1989.

[HC93] Q. Huang and T.A. Cruse. Some notes on singular integral techniques in boundary element analysis. *Int. J. Numer. Meth. Engrg.*, 36:2643–2659, 1993.

[HC94] Q. Huang and T.A. Cruse. On the non-singular traction-BIE in elasticity. *Int. J. Numer. Meth. Engrg.*, 37:2041–2072, 1994.

[Her97] U. Herding. *Eine Field-Boundary-Element-Formulierung für elastoplastische Stoffmodelle der Schädigungsmechanik.* PhD thesis, Universität Erlangen-Nürnberg, 1997.

[Hil94] J.O. Hildenbrand. *Hypersinguläre Integralgleichungen in der Randelementmethode - Berechnung der Integrale und Anwendung bei linearelastischen Aufgaben.* Fortschritt-Berichte, Reihe 18, Nr. 143, VDI-Verlag, Universität Erlangen-Nürnberg, Düsseldorf, 1994.

[HN89] W. Hackbusch and Z.P. Nowak. On the fast matrix multiplication in the boundary element method by panel clustering. *Numer. Math.*, 54:463–491, 1989.

[Hol92] S. M. Holzer. *Das symmetrische Randelementeverfahren: Numerische Realisierung und Kopplung mit der Finite-Elemente-Methode zur elastoplastischen Strukturanalyse.* PhD thesis, Technische Universität München, Munich, 1992.

[Hör90] L. Hörmander. *The Analysis of Linear Partial Differential Operators*, volume I. Springer, New York, 2nd edition, 1990.

[Hue96] A. Huesmann. *Behandlung des reibbehafteten elastoplastischen Kontakts mittels der Randelementmethode.* VDI Fortschritt–Berichte 18/187, Universität Erlangen-Nürnberg, VDI-Verlag, Düsseldorf, 1996.

[Jah98] P. Jahn. *Die Randelement-Methode für elastisch gebettete Platten.* PhD thesis, Universitäts-Gesamthochschule Kassel, 1998.

[Jia89] Y.S. Jiang. *Slope Analysis Using Boundary Elements.* Number 52 in Lecture Notes in Engineering. Springer, Berlin, 1989.

[Kai94] G. Kaiser. *A Friendly Guide to Wavelets.* Birkhäuser, Basel, 1994.

[KB93] J.H. Kane and C. Balakrishna. Symmetric Galerkin boundary formulations employing curved elements. *Int. J. Numer. Meth. Engrg.*, 36:2157–2187, 1993.

[KG00] M. Kögl and L. Gaul. A 3-D boundary element method for dynamic analysis of anisotropic elastic solids. *Comput. Model. Engrg. Sci.*, 1(4):27–44, 2000.

[Kie91] R. Kieser. *Über einseitige Sprungrelationen und hypersinguläre Operatoren in der Methode der Randelemente.* PhD thesis, Universität Stuttgart, 1991.

[Kit85] M. Kitahara. *Boundary Integral Equation Methods in Eigenvalue Problems of Elastodynamics and Thin Plates.* Elsevier, Amsterdam, 1985.

[Kle90] P. Klein. *Zur Beschreibung der dynamischen Wechselwirkung von Fundamentstrukturen mit dem viskoelastischen Baugrund durch dreidimensionale Randelementformulierungen.* PhD thesis, Universität der Bundeswehr Hamburg, 1990.

[Kög00] M. Kögl. *A Boundary Element Method for Dynamic Analysis of Anisotropic Elastic, Piezoelectric, and Thermoelastic Solids.* PhD thesis, Universität Stuttgart, 2000.

[Kre92] K. Kremer. *Randelement-Methode für elasto-plastische Strukturberechnungen im Grundbau und ihre Kopplung mit der Finite-Element-Methode.* PhD thesis, Technische Universität Braunschweig, 1992.

[Kre96] M. Kreienmeyer. *Kopplung von Rand- und Finite-Element-Methoden für ebene Elastoplastizität mit Implementierung auf Parallelrechnern.* PhD thesis, Universität Hannover, 1996.

[Krö53] E. Kröner. Das Fundamentalintegral der anisotropen elastischen Differentialgleichungen. *Z. Physik*, 136:402–410, 1953.

[KRR92] G. Krishnasamy, F.J. Rizzo, and T.J. Rudolphi. Continuity requirements for density functions in the boundary integral equation method. *Comput. Mech.*, 9:267–284, 1992.

[Kun01] A. Kunoth. *Wavelet Methods - Elliptic Boundary Value Problems and Control Problems.* Teubner, Stuttgart, 2001.

[Kyt96] P.K. Kythe. *Fundamental Solutions for Differential Operators and Applications.* Birkhäuser, Boston, 1996.

[Lac75] J.C. Lachat. *A Further Development of the Boundary Integral Technique for Elastostatics.* PhD thesis, University of Southhampton, 1975.

[Lag95] C. Lage. *Softwareentwicklung zur Randelementmethode: Analyse und Entwurf effizienter Techniken.* PhD thesis, Christian-Albrechts-Universität Kiel, 1995.

[LC90] J. Lemaitre and J.-L. Chaboche. *Mechanics of Solid Materials.* Cambridge University Press, 1990.

[Leh97] R. Lehmann. *Studies on the Use of the Boundary Element Method in Physical Geodesy.* Verlag der Bayerischen Akademie der Wissenschaften, Munich, 1997.

[Lei94] V.M.A. Leitão. *Boundary Elements in Nonlinear Fracture Mechanics.* Computational Mechanics Publications, Southhampton, 1994.

[Lek68] S.G. Lekhnitskii. *Anisotropic Plates.* Gordon & Breach, New York, 1968.

[LG99] A. Lorenzana and J.A. Garrido. Analysis of the elastic-plastic problem involving finite plastic strain using the boundary element method. *Comput. Struct.*, 73:147–159, 1999.

[Lie97] M. Lieb. *Adaptive numerische Fouriertransformation in der Bodendynamik unter Verwendung einer Waveletzerlegung.* PhD thesis, Technische Universität München, Munich, 1997.

[Lig66] M.J. Lighthill. *Einführung in die Theorie der Fourieranalysis und der Theorie der verallgemeinerten Funktionen.* Bibliographisches Institut, Mannheim, 1966.

[LL83] J.A. Liggett and P.L.F. Liu. *The Boundary Integral Equation Method for Porous Media Flow.* Allen & Unwin, London, 1983.

[LM72] J. L. Lions and E. Magenes. *Non-Homogeneous Boundary Value Problems and Applications*, volume I. Springer, Berlin, 1972.

[LM73] J. L. Lions and E. Magenes. *Non-Homogeneous Boundary Value Problems and Applications*, volume III. Springer, Berlin, 1973.

[LMR94] A.K. Louis, P. Maaß, and A. Rieder. *Wavelets.* Teubner, Stuttgart, 1994.

[LS97] C. Lage and C. Schwab. Wavelet Galerkin algorithms for boundary integral equations. Preprint, Seminar für Angewandte Mathematik, ETH Zurich, 1997.

[Mai97] M. Maischak. The analytical computation of the Galerkin elements for the Laplace, Lamé, and Helmholtz equation in 3D-BEM. Preprint, Universität Hannover, 1997.

[Mai99] M. Maischak. The analytical computation of the Galerkin elements for the Laplace, Lamé, and Helmholtz equation in 2D-BEM. Preprint, Universität Hannover, 1999.

[Mal89] S. Mallat. A theory for multiresolution signal decomposition: The wavelet representation. In *IEEE Trans. Pattern Anal. Machine Intell.*, volume 11, pages 674–693, 1989.

[MB88] G.D. Manolis and D.E. Beskos. *Boundary Element Methods in Elastodynamics.* Unwin Hyman, London, 1988.

[McL00] W. McLean. *Strongly Elliptic Systems and Boundary Integral Equations.* Cambridge University Press, 2000.

[MD92] G.D. Manolis and T.G. Davies, editors. *Boundary Element Techniques in Geomechanics.* Computational Mechanics Publications, Southhampton, 1992.

[MDC89] G. Maier, M. Diligenti, and A. Carini. A vatiational formulation of the boundary element method in elastodynamics. In B.S. Annigeri and K. Tseng, editors, *Boundary Element Methods in Engineering*, pages 268–275, Berlin, 1989. Springer.

[MEIL01] N.S. Mera, L. Elliott, D.B. Ingham, and D. Lesnic. A comparison of boundary element method formulations for steady state anisotropic heat conduction problems. *Engrg. Anal. Bound. Elem.*, 25(2):115–128, 2001.

[Mey90] Y. Meyer. *Ondelettes et Opérateurs I - Ondelettes.* Hermann, Paris, 1990.

[MMP95] S.G. Mikhlin, N.F. Morozov, and M.V. Paukshto. *The Integral Equations of the Theory of Elasticity*, volume 135 of *Teubner-Texte zur Mathematik.* Teubner, Stuttgart, 1995.

[MMS97] M. Maischak, P. Mund, and E.P. Stephan. Adaptive multilevel BEM for acoustic scattering. Preprint, Universität Hannover, 1997.

[MP86] S.G. Mikhlin and S. Prößdorf. *Singular integral operators.* Springer, Berlin, 1986.

[MP96] W. McLean and S.B. Prößdorf. Boundary element collocation methods using splines with multiple knots. *Numer. Math.*, 74:419–451, 1996.

[MP98] V. Mantič and F. París. Integral kernels in the 2D Somigliana displacement and stress identities for anisotropic materials. *Comput. Mech.*, 22:77–87, 1998.

[MR96] P.A. Martin and F.J. Rizzo. Hypersingular integrals: how smooth must the density be? *Int. J. Numer. Meth. Engrg.*, 39:687–704, 1996.

[Muk77] S. Mukherjee. Corrected boundary integral equations in planar thermoplasticity. *Int. J. Solids Struct.*, 13:331–335, 1977.

[Muk82] S. Mukherjee. *Boundary Element Methods in Creep and Fracture.* Elsevier, London, 1982.

[Mun97] P. Mund. *Zwei-Level-Verfahren für Randintegralgleichungen mit Anwendungen auf die nichtlineare FEM-BEM-Kopplung.* PhD thesis, Universität Hannover, 1997.

[Mus92] N.I. Muskhelishvili. *Singular Integral Equations.* Dover Publications, Inc., New York, 2nd edition, 1992.

[MW96] L. Morino and W.L. Wendland, editors. *IABEM Symposium on Boundary Integral Methods for Nonlinear Problems.* Kluwer, Dordrecht, 1996.

[NB82] G. Novati and C.A. Brebbia. Boundary element formulation for geometrically non-linear elastostatics. *Appl. Math. Modelling*, 6:136–8, 1982.

[NBN91] J.-C. Nedelec, E. Becache, and N. Nishimura. Regularization in 3D for anisotropic elastodynamic crack and obstacle problems. In L. Morino and R. Piva, editors, *Boundary Integral Methods - Theory and Applications*, Proceedings of the IABEM Symposium, Rome, Italy, October 15–19, 1990, pages 379–389, Berlin, 1991. Springer.

[Ned82] J.-C. Nedelec. Integral equations with non-integrable kernels. *Int. Eq. Oper. Theor.*, 5:563–572, 1982.

[Nei94] T. Neidhart. *Lösung dreidimensionaler linearer Probleme der Bodendynamik mit der Randelementmethode.* PhD thesis, Universität Fridericiana Karlsruhe, 1994.

[Neu82] W. Neureiter. *Boundary-Element-Programmrealisierung zur Lösung von zwei- und dreidimensionalen Thermoelastischen Problemen mit Volumenkräften.* PhD thesis, Technische Universität München, 1982.

[Nis88] N. Nishimura. *Applications of the Boundary Integral Equation Methods to Solid Mechanics.* PhD thesis, Kyoto University, 1988.

[Nor94] A.N. Norris. Dynamic Green's functions in anisotropic piezoelectric, thermoelastic, and poroelastic solids. *Proc. Royal Soc. London A*, 447:175–188, 1994.

[Now75] W. Nowinski. *Dynamic problems of thermoelasticity.* Noordhoff Int. Publ., Leyden, 1975.

[Obe92] M. Oberguggenberger. *Multiplication of Distributions and Applications to Partial Differential Equations.* Longman Scientific and Technical, 1992.

[OR94] M. Oberguggenberger and E.E. Rosinger. *Solution of Continuous Nonlinear PDEs through Order Completion.* North-Holland, Amsterdam, 1994.

[PA96] E. Pan and B. Amadei. Fracture mechanics analysis of cracked 2-D anisotropic media with a new formulation of the boundary element method. *Int. J. Fract.*, 77:161–174, 1996.

[PBW92] P.W. Partridge, C.A. Brebbia, and L.C. Wrobel. *The Dual Reciprocity Boundary Element Method.* Computational Mechanics Publications, Southhampton, 1992.

[PK99] M.N. Perelmuter and S.V. Kuznetsov. BEM analysis of 3D stress concentration problems of anisotropic composite structures. In W. Wunderlich, editor, *Europ. Conf. Comput. Mech. (ECCM)*, Munich, 1999.

[Pol96] R. Pollandt. *Ein Approximationsverfahren zur Lösung nichtlinearer Differentialgleichungen auf Grundlage der Boundary Element Methode und von Radialbasisfunktionen.* PhD thesis, Hochschule für Architektur und Bauwesen, Weimar, 1996.

[Pom98] A. Pomp. *The Boundary-Domain Integral Method for Elliptic Systems*, volume 1683 of *Lecture Notes in Mathematics*. Springer, Berlin, 1998.

[Pot87] I. Potrc. *Zur Behandlung temperaturabhängiger elastoplastischer Probleme mittels Randelementmethode.* PhD thesis, Universität Erlangen-Nürnberg, 1987.

[Poz92] C. Pozrikidis. *Boundary Integral and Singularity Methods for Linearized Viscous Flow.* Cambridge University Press, 1992.

[Pro85] B. Protopsaltis. *Rand-Integralgleichungen für dünne Platten und ihre Kopplung mit finiten Elementen.* PhD thesis, Technische Universität München, Munich, 1985.

[PS97] T.v. Petersdorff and C. Schwab. *Wavelet Analysis and its Application*, volume 6, chapter Fully discrete multiscale Galerkin BEM, pages 287–346. Academic Press, New York, 1997.

[PSS97] T.v. Petersdorff, C. Schwab, and R. Schneider. Multiwavelets for second-kind integral equations. *SIAM - J. Numer. Anal.*, 34(6):2212–2227, 1997.

[PY00] E. Pan and F.G. Yuan. Boundary element analysis of three-dimensional cracks in anisotropic solids. *Int. J. Numer. Meth. Engrg.*, 48:211–237, 2000.

[Ran85] E. Rank. *A-posteriori-Fehlerabschätzungen und adaptive Netzverfeinerung für Finite-Element- und Randintegralelement-Methoden.* PhD thesis, Technische Universität München, Munich, 1985.

[Ras00] Y.F. Rashed. *Boundary element formulations for thick plates.* Computational Mechanics Publications, Southhampton, 2000.

[Rat96a] A. Rathsfeld. A wavelet algorithm for the boundary element solution of a geodetic boundary value problem. Technical report, Weierstraß-Institut für Angewandte Analysis und Stochastik, Berlin, 1996.

[Rat96b] A. Rathsfeld. A wavelet algorithm for the solution of a singular integral equation over a smooth two-dimensional manifold. Technical report, Weierstraß-Institut für Angewandte Analysis und Stochastik, Berlin, 1996.

[Rat97] A. Rathsfeld. On the stability of piecewise linear wavelet collocation and the solution of the double layer equation over polygonal curves. Technical report, Weierstraß-Institut für Angewandte Analysis und Stochastik, Berlin, 1997.

[RC99] J.D. Richardson and T.A. Cruse. Weakly singular stress-BEM for 2D elastostatics. *Int. J. Numer. Meth. Engrg.*, 45:13–35, 1999.

[Ric73] P.C. Riccardella. *An implementation of the boundary-integral-technique for planar problems in elasticity and elastoplasticity.* PhD thesis, Pittsburgh, Pennsylvania, 1973.

[Ric96] C. Richter. *Eine Greensche Funktion der Halbscheibe und ihre Einbindung in die Zeitbereichs-Randelementmethode.* PhD thesis, Ruhr-Universität Bochum, 1996.

[Rok85] V. Rokhlin. Rapid solution of integral equations of classical potential theory. *J. Comput. Phys.*, 60:187–207, 1985.

[Ruß92] S. Rußwurm. *Berechnung elastoplastischer Rißprobleme mittels der Randelementmethode.* Fortschritt-Berichte, Reihe 18, Nr. 104, Universität Erlangen-Nürnberg, VDI-Verlag, Düsseldorf, 1992.

[RW91] R.K.N.D. Rajapakse and Y. Wang. Elastodynamic Green's functions of orthotropic half plane. *ASCE - J. Engrg. Mech.*, 117(3):588–604, 1991.

[Sal97] A.L. Saleh. *Crack Growth in Concrete Using Boundary Elements.* Computational Mechanics Publications, Southhampton, 1997.

[Sal98] J. Salençon. *Mécanique des milileux continus*, volume I-III. Ecole Polytechnique, Palaiseau, 1998.

[Sar89] B. Sarler. *Contributions in Using Boundary Element Methods for Calculation of Nonlinear Axialsymmetric Transient Heat Diffusion Equation in Deformable Geometry.* PhD thesis, University of Maribor, Slovenia, 1989.

[Sau92] S.A. Sauter. *Über die effiziente Verwendung des Galerkinverfahrens zur Lösung Fredholmscher Integralgleichungen.* PhD thesis, Universität Kiel, 1992.

[SB89] G. Shi and G. Bézine. The direct boundary integral equation method for the free vibration analysis of orthotropic plates. *Eur. J. Mech. A/Solids*, 8:277–291, 1989.

[SC71] J.L. Swedlow and T.A. Cruse. Formulation of boundary integral equations for three-dimensional elastoplastic flow. *Int. J. Solids Struct.*, 7:1673–1681, 1971.

[Sch54] L. Schwartz. Sur l'impossibilité de la multiplication des distributions. *C. R. Acad. Sci.*, 239:847–848, 1954.

[Sch94a] M. Schanz. *Eine Randelementformulierung im Zeitbereich mit verallgemeinerten viskoelastischen Stoffgesetzen.* PhD thesis, Universität der Bundeswehr Hamburg, 1994.

[Sch94b] N.A. Schclar. *Anisotropic Analysis using Boundary Elements.* Computational Mechanics Publications, Southhampton, 1994.

[Sch98] R. Schneider. *Multiskalen- und Wavelet-Matrixkompression.* Teubner, Stuttgart, 1998.

[Sch51] L. Schwartz. *Théorie des distributions*, volume I-II. Hermann & Cie, Paris, 1950/51.

[SD01] A. Saez and J. Dominguez. Dynamic crack problems in three-dimensional transversely isotropic solids. *Engrg. Anal. Bound. Elem.*, 25(3):203–210, 2001.

[SG91] P.D. Spanos and R. Ghanem. Boundary element formulation for random vibration problems. *ASCE - J. Engrg. Mech.*, 117(2):409–423, 1991.

[SG98] M.A. Sales and L.J. Gray. Evaluation of the anisotropic Green's function and its derivatives. *Comput. Struct.*, 69:247–254, 1998.

[Sha93] R.P. Shaw. A history of boundary elements. In C.A. Brebbia and J.J. Rencis, editors, *Boundary Elements XV - Fluid Flow and Computational Aspects*, volume 1, pages 265–280, Southhampton, 1993. Computational Mechanics Publications.

[Sic89] W. Sichert. *Berechnung von instationären thermischen Problemen mittels der Randelementmethode.* PhD thesis, Universität Erlangen-Nürnberg, 1989.

[Sip88] H. Sippel. *Elastoplastische Berechnungen mittels der Boundary-Element-Methode.* PhD thesis, Universität Erlangen-Nürnberg, 1988.

[Slo92] I. H. Sloan. Error analysis for boundary integral methods. *Acta Numerica*, 1:287–339, 1992.

[SMNM92] S. Sirtori, G. Maier, G. Novati, and S. Miccoli. A Galerkin symmetric boundary element method in elasticity: formulation and implementation. *Int. J. Numer. Meth. Engrg.*, 35:255–282, 1992.

[SS82] J. Sládek and V. Sládek. Three-dimensional crack-analysis for an anisotropic body. *Appl. Math. Modelling*, 6:374–380, 1982.

[SS95] J. Sládek and V. Sládek. Boundary element method for thermoelastoplastic problems. *Int. J. Numer. Meth. Engrg.*, 38(21):3635–3652, 1995.

[SS00] G. Schmidlin and C. Schwab. Wavelet Galerkin BEM on unstructured meshes by aggregation. Research report 2000-15, Seminar für Angewandte Mathematik, ETH Zürich, Zurich, 2000.

[SSLV98] V. Sládek, J. Sládek, and A. Le Van. Completely regularized integral representations and integral equations for anisotropic bodies with initial strains. *Z. Angewandte Math. Mech.*, 78(11):771–780, 1998.

[ST99] Y.C. Shiah and C.L. Tan. Exact boundary integral transformation of the thermoelastic domain integral in BEM for general 2D anisotropic elasticity. *Comput. Mech.*, 23:87–96, 1999.

[Sta79] I. Stakgold. *Green's Functions and Boundary Problems.* John Wiley & Sons, New York, 1979.

[Sta00] G.E. Stavroulakis. *Inverse and Crack Identification Problems in Engineering Mechanics.* Kluwer, Dordrecht, 2000.

[Ste84] E.P. Stephan. *Boundary Integral Equations for Mixed Boundary Value Problems, Screen and Transmission Problems in* $\mathbb{R}^3$. Habilitation, Technische Hochschule Darmstadt, 1984.

[Ste93] B. Steinfeld. *Numerische Berechnung dreidimensionaler Kontaktprobleme Bauwerk-Boden mittels zeitabhängiger Randintegralgleichungen der Elastodynamik.* PhD thesis, Ruhr-Universität Bochum, 1993.

[Ste95] R. Steinbauer. Colombeau-theorie und ultrarelativistischer Limes. Master's thesis, Universität Wien, 1995.

[Ste00] O. Steinbach. On a hybrid boundary element method. *Numer. Math.*, 84:679–695, 2000.

[STW90] A. H. Schatz, V. Thomée, and W. L. Wendland. *Mathematical Theory of Finite and Boundary Element Methods.* Birkhäuser, Basel, 1990.

[SW88] E. Stein and W. Wendland, editors. *Finite Element and Boundary Element Techniques from Mathematical and Engineering Point of View.* Number 301 in Int. Center Mech. Studies (CISM). Springer, New York, 1988.

[Swe98] W. Sweldens. The lifting scheme: a construction of second generation wavelets. *SIAM J. Math. Anal.*, 29(2):511–546, 1998.

[Tak91] S. Takahashi. *Elastic Contact Analysis by Boundary Elements.* Springer, New York, 1991.

[Tam92] A. Tamborek. *Erschütterunsausbreitung vom Rad/Schiene-System bei Damm, Einschnitt und Ebene.* PhD thesis, Universität Fridericiana Karlsruhe, 1992.

[Tel83] J.F.C. Telles. *The Boundary Element Method Applied to Inelastic Problems.* Springer, Berlin, 1983.

[TG92] C.L. Tan and Y.L. Gao. Boundary integral equation fracture mechanics analysis for plane orthotropic bodies. *Int. J. Fract.*, 53:343–365, 1992.

[Tim53] S.P. Timoshenko. *History of Strength of Materials.* McGraw-Hill, New York, 1953.

[Tre94] J. Trevelyan. *Boundary Elements for Engineers : Theory and Applications.* Computational Mechanics Publications, Southhampton, 1994.

[Tri89] T. Triantafyllidis. *Halbraumlösungen zur Behandlung bodendynamischer Probleme mit der Randelementmethode.* Habilitation, Universität Fridericiana Karlsruhe, 1989.

[Trö95] G. Tröndle. *Effiziente Schallberechnung mit einem adaptiven Mehrgitterverfahren für die 3-D Randelementmethode.* PhD thesis, Technische Universität Braunschweig, 1995.

[TSS94] M. Tanaka, V. Sládek, and J. Sládek. Regularization techniques applied to boundary element methods. *ASME - Appl. Mech. Review*, 47:457–499, 1994.

[Ven83] W.S. Venturini. *Boundary Element Method in Geomechanics.* Number 4 in Lecture Notes in Engineering. Springer, Berlin, 1983.

[VR73] S.M. Vogel and F.J. Rizzo. An integral equation formulation of three dimensional anisotropic elastostatic boundary value problems. *J. Elast.*, 3(3):203–216, 1973.

[WA93] C.Y. Wang and J.D. Achenbach. A new method to obtain 3-D Green's functions for anisotropic solids. *Wave Motion*, 18:273–289, 1993.

[WA94] C.Y. Wang and J.D. Achenbach. Elastodynamic fundamental solutions for anisotropic solids. *Geophys. J. Int.*, 118:384–392, 1994.

[WA95] C.Y. Wang and J.D. Achenbach. Three-dimensional time-harmonic elastodynamic Green's functions for anisotropic solids. *Proc. Royal Soc. London A*, 449:441–458, 1995.

[Wag00] N. Wagner. *Untersuchung der Boden-Fahrzeug-Interaktion mit gekoppelter Rand- und Finite-Element-Methoden.* PhD thesis, Technische Universität Braunschweig, 2000.

[Wan97a] C.Y. Wang. Elastic fields produced by a point source in solids of general anisotropy. *J. Engrg. Math.*, 32(1):41–52, 1997.

[Wan97b] G. Wang. Application of wavelets on the interval to numerical analysis of integral equations in electromagnetic scattering problems. *Int. J. Numer. Meth. Engrg.*, 40:1–13, 1997.

[Wat97] Waterloo Maple Software, University of Waterloo. *Maple V Release 5*, 1997.

[WC78] R.B. Wilson and T.A. Cruse. Efficient implementation of anisotropic three dimensional boundary integral equation stress analysis. *Int. J. Numer. Meth. Engrg.*, 12:1383–1397, 1978.

[Wen97] W.L. Wendland, editor. *Boundary Element Topics*. Springer, Berlin, 1997. Proceedings of the Final Conference of the Priority Research Progamme "Boundary Element Methods 1989–1995" of the German Research Foundation, October 2–4, 1995 in Stuttgart.

[Wie90] L. Wienert. *Die numerische Approximation von Randintegraloperatoren für die Helmholtzgleichung im* $\mathbb{R}^3$. PhD thesis, Universität von Göttingen, 1990.

[Wie93] T. Wiebe. *Wellenausbreitung in poroelastischen Medien: Untersuchung mit Randintegralgleichungen*. PhD thesis, Technische Universität Braunschweig, 1993.

[WJT00] W. Wang, X. Ji, and M. Tanaka. A dual reciprocity boundary element approach for the problems of large deflection of thin elastic plates. *Comput. Mech.*, 26:58–65, 2000.

[WNB86] L.C. Wrobel, D. Nardini, and C.A. Brebbia. *The Dual Reciprocity Boundary Element Formulation for Transient Heat Conduction*, volume 6 of *Finite Elements in Water Ressources*. Springer, Berlin, 1986.

[Wol79] K.B. Wolf. *Integral Transforms in Science and Engineering*. Plenum Press, New York, 1979.

[You88] Y. Youn. *Determination of effective elastic properties and thermal residual stresses in fiber-reinforced composites by the boundary element method*. PhD thesis, Michigan State University, 1988.

[Zha91] Z. Zhao. *Shape Design Sensitivity and Optimization Using Boundary Element Method*. Springer, New York, 1991.

[Zot85] R. Zotemantel. *Berechnung von Platten nach der Methode der Randelemente*. PhD thesis, Universtiät Dortmund, 1985.

[ZTA97] J.J. Zhang, C.L. Tan, and F.F. Afagh. Treatment of the body-force volume integrals in BEM by exact transformation for 2-D anisotropic elasticity. *Int. J. Numer. Meth. Engrg.*, 40:89–109, 1997.

GPSR Compliance

The European Union's (EU) General Product Safety Regulation (GPSR) is a set of rules that requires consumer products to be safe and our obligations to ensure this.

If you have any concerns about our products, you can contact us on ProductSafety@springernature.com

In case Publisher is established outside the EU, the EU authorized representative is:

Springer Nature Customer Service Center GmbH
Europaplatz 3
69115 Heidelberg, Germany

Batch number: 10584448

Printed by Printforce, the Netherlands